Invasive Technification

ALSO AVAILABLE FROM BLOOMSBURY

Intensive Science & Virtual Philosophy, Manuel DeLanda
Nanoethics, Donal P. O'Mathuna
A New Philosophy of Society, Manuel DeLanda
Philosophy and Simulation, Manuel DeLanda

Invasive Technification

Critical Essays in the Philosophy of Technology

Gernot Böhme

Translated by Cameron Shingleton

B L O O M S B U R Y
LONDON • NEW DELHI • NEW YORK • SYDNEY

Bloomsbury Academic
An imprint of Bloomsbury Publishing Plc

50 Bedford Square	175 Fifth Avenue
London	New York
WC1B 3DP	NY 10010
UK	USA

www.bloomsbury.com

Originally published in German as *Invasive Technisierung: Technikphilosophie une Technikkritik*
© Die Graue Edition, 2008

This translation © Bloomsbury Publishing Plc, 2012

British Library Cataloguing-in-Publication Data
A catalogue record for this book is available from the British Library.

ISBN: HB: 978-1-4411-4901-5
PB: 978-1-4411-8294-4

Library of Congress Cataloging-in-Publication Data
Böhme, Gernot.
[Invasive Technisierung. English]
Invasive technification: critical essays in the philosophy of technology/Gernot Böhme;
translated by Cameron Shingleton. p. cm.
Includes bibliographical references and index.
ISBN 978-1-4411-4901-5 (hardcover: alk. paper) – ISBN 978-1-4411-8294-4 (pbk.: alk. paper) – ISBN 978-1-4411-3402-8
(ebook epub: alk. paper) – ISBN 978-1-4411-9465-7 (ebook pdf: alk. paper) 1. Technology – Philosophy. I. Title.

T14.B538 2012
601–dc23
2012001745

Typeset by Deanta Global Publishing Services, Chennai, India
Printed and bound in Great Britain

CONTENTS

PREFACE

The present volume brings together work from many decades of reflection about technology and technological development. The essays and papers it contains are consistently critical. They aim to spell out the meaning of technological development for human life in a range of major contexts.

My work in this field, some of which would count formally as part of the history of technology or the sociology of technology or the philosophy or ethics of technology, was given considerable stimulus by the post-graduate seminar series with the general title 'Technification and Society' held at Darmstadt University. Co-founded in 1997 by myself and several colleagues, the programme ran until 2006. Our project was funded by the German Research Council. At its heart was an inter-disciplinary attempt to study the relationship between technology and society in terms of what we called processes of technification.[1]

My thanks go to all colleagues who were part of this decade-long intellectual labour for years of fruitful discussion. This book represents my contribution to the work of those interesting years. My hope is that it will not only stimulate discussion within the philosophy of technology and in specialist academic circles, but also among a broader reading public. If the book can in some way help with the task of orienting ourselves within technological civilization then it will have well and truly served its purpose. It was written in the conviction that taking stock of where technological developments have brought us is one of the most urgent tasks of contemporary thought.

---

[1] For references to individual projects and dissertations associated with the programme, see
http://www.ifs.tu-darmstadt.de/fileadmin/gradkoll/index.html
For group work that the programme produced, the reader is referred to Böhme, Manzei (Eds.), *Kritische Theorie der Technik und der Natur*. Munich: Fink, 2003 and Heil, Kaminski, Stippak, Unger (Eds.), *Tensions and Convergences. Technological and Aesthetic Transformations of Society*. Bielefeld: transcript 2007, pp. 23–35.

1

Introduction

Invasive technification

Deep Within the Brain

Not long ago, a book about an illness achieved the notable feat of making it onto German bestseller lists. The interesting question is – why? Could it have had something to do with the way it managed to move the reader to participate in the story? There are, after all, a plethora of medical narratives that aim to mobilize or strengthen our affective powers. Could it have been simply that this particular account was well written, that it displayed an authorial frankness, placing it squarely in the tradition of European confessional literature? My suspicion is that the breadth and depth of the public response had to do more with the content of the account – a content that triggers a sort of metaphysical terror in the reader.

The book is called *Deep Within the Brain* and in it the sociologist Helmut Dubiel relates the tale of his experiences with Parkinson's disease. The decisive point of the narrative – what makes it stand out from many a biographical account of illness – is when the progress of Dubiel's malady can no longer be kept in check with medication and he is coming to terms with public humiliations because of ever more noticeable symptoms; at which stage he has a technically state-of-the-art implant inserted into his brain. The implant creates an electrical stimulus that regulates the interaction of brain hormones.[1] (The latter become unbalanced in the absence of dopamine, which is the root cause of Parkinson's disease.) Yet although artificial regulation does indeed result in a notable improvement in typical Parkinson's symptoms, such as shaking and reduced motor functions, in some cases, like Dubiel's, it has extremely severe side effects. The stimulus to the interior of the brain seriously interferes with the speech capacity of the patient – a catastrophic impairment for an academic with a lecturing position at a university.

Thus, Dubiel could not but view the medical intervention subjectively as a mistake and objectively as a failure. In his case, the medical world had hit on a form of treatment that in a certain sense seems macabre. As Dubiel reported in a newspaper interview, to other people he sometimes comes across as a sort of zombie: using a handheld remote control, he can switch off the implanted pacemaker that stimulates the relevant region deep in the brain, allowing him to speak again without any trouble; however, his Parkinson's symptoms recur, with the result that he can no longer walk. When he switches the device on again, he can walk and carry out deliberate bodily movements, but his speech capacity is again impaired – so much so that he can no longer give a lecture.

Dubiel's case provides dramatic proof of where society stands today vis-à-vis the new possibilities of technified medical treatment. When Dubiel talks about being able to switch *himself* on and off, what is fundamentally at issue is the meaning of this selfhood. Is he the human being who can move about in well co-ordinated fashion or the human being who can form proper sentences? Or is he perhaps the human being who has the subjective experience of switching on and off? The implant seems to have brought about a deep-seated dissociation of Dubiel's psychophysical unity. However, the medical technology, already viewed by medical practitioners as *invasive* – that is, as intruding deeply and permanently into the body rather than merely supporting or supplementing bodily functions – can also be called invasive in a wider sense. In cases like Dubiel's, it has failed to restore a human life damaged by illness or to allow such a life to be lived halfway normally again, and instead changed what life fundamentally is for the patient – changed, one could even say, *what he is as a human being*.

Today especially, developments are afoot in medicine that give good reason to critically apply the medical concept of *the invasive* to the technification of human relationships and of social relations across the board. It is true that the concept has military connotations – an invasion is, of course, a violent act of intrusion. Much medical technology, however, has to be viewed in the same light, indeed it is already viewed as such by the patients affected by it,[2] and even sensed *bodily* as such, in the depths of our organic being.[3]

The decisive question though is whether technification merely *improves and extends* an already existing form of human action or already existing human relations, or whether it fundamentally *alters* them. Clearly, it is possible to object that technologies have always brought about fundamental changes to human life. Thus, with the invention of writing (a communication technology), human society, and indeed what it was to be human, was fundamentally changed.[4] Yet obviously, until now, we have had little reason to see technological change in the problematic light we do today. Technology has, instead, been viewed almost totally from the point of view of means and ends – a way of looking at things that takes the ends wholly for granted. The result is that the ethics of technology, including medical ethics, have been almost exclusively utilitarian; problematic situations have been thought to demand a weighing up of costs versus benefits or expectations versus risks. What has been overlooked is that the application of a new technology can change the preconditions of the application, and that means also the very purposes and ends of the application. The utilitarian approach to technology no longer suffices, for example in light of the experiences medical technology confronts us with today. In general the question has to be asked: what does technology mean when it takes the form of invasive technification of the most basic conditions of human life?

The iron cage

At the beginning of the twentieth century, the sociologist Max Weber famously asserted that modern man was in the process of confining himself to an iron cage. What he had in mind were the constraints that human beings have to impose on themselves to make the rationalization of social life possible – at work and on the street in traffic, in business and in private life. A characteristic example is punctuality, a virtue without which the worlds of modern transport and work would simply be unthinkable. According to Weber, the detachment vis-à-vis one's own feelings and moods that is necessary to operate technical equipment is radical. Human behaviour necessarily becomes cooler, more matter of fact.

By technology or technique (*Technik*) Weber understands the thoroughgoing rational organization of activity for the purposes of increased efficiency. In this sense, we speak not only of power generation technology, but also of techniques of playing the piano. The concept of technology/technique involved here implies that the rationalization of the conditions of human life is a kind of technification. For Weber, feats of individual moral self-control form the background to the efficient rational arrangement of social life, above all feats of inner discipline, of worldly asceticism, such as industriousness, reliability, punctuality and matter-of-factness; in short, the prime values associated with the puritan ethic that made the modern world possible. So for Weber the problem of the iron cage is the social problem of *technical efficiency* on the one hand and its *ethical preconditions* on the other. In 1954, Jacques Ellul drew together both sides of Weber's analysis under the rubric of *technological society*.[5]

Today, the Weberian metaphor of the iron cage makes a curious impression. For Weber is not thinking of any physical structure, say the rail network that in his day was imprinting a new spatial order on the landscape; nor is he thinking of the steel and reinforced concrete structures that became the physical home of working and business life and, in the form of warehouses, helped to transform luxury consumption into mass consumption. Quite the opposite: the iron cage is, for Weber, precisely a metaphor for an inner state, for inner ethical constraints or at most abstract social constraints.

A generation after Weber, the growth of such constraints since early modern times was described by another sociologist, Norbert Elias, as *the civilizing process*, in a work with precisely that title.[6] The notable point about Elias' study, however, is that in the very period of its inception, viz. the 1930s, the process itself was veering round 180 degrees – something that is easy to gauge in the sole passage where Elias addresses the issue of modern technology. It is worth citing in full:

> A simple way of picturing the difference between the integration of the individual within a complex society and within a less complex one is to think of their different road systems. These are in a sense spatial functions of a social integration which, in its totality, cannot be expressed merely in terms of concepts derived from the four-dimensional continuum. One should think of the country roads of a simple warrior society with a barter economy, uneven, unmetalled, exposed to damage from wind and rain. With few exceptions, there is very little traffic; the main danger which man here represents for other men is an attack by soldiers or thieves. When people look around them, scanning the trees and hills or the road itself, they do so primarily because they must always be prepared for armed attack, and only secondarily because they have to avoid collision. Life on the main roads of this society demands a constant readiness to fight, and free play of the emotions in defence of one's life or possessions from physical attack. Traffic on the main roads of a big city in the complex society of our time demands a quite different moulding of the psychological apparatus. Here the danger of physical attack is minimal. Cars are rushing in all directions; pedestrians and cyclists are trying to thread their way through the *mêlée* of cars; policemen stand at the main crossroads to regulate the traffic with varying success. But this external control is founded on the assumption that every individual is himself regulating his behaviour with the utmost exactitude in accordance with the necessities of this network. The chief danger that people here represent for others results from someone in this bustle losing his self-control. A constant and highly differentiated regulation of one's own behaviour is needed for the individual to steer his way through traffic. If the strain of such constant self-control becomes too much for an individual, this is enough to put himself and others in mortal danger.[7]

Elias, of course, makes the comparison between the street in a traditional warrior society and the street of a modern metropolis in order to illustrate his thesis that the move from external constraint to self-constraint plays a key role in shaping the behaviour of individuals in transit and the sort of attention that they pay to potential dangers. However, because he describes a scene involving *technologically regulated* traffic, it becomes clear that in twentieth-century modernity a shift *back to external constraints* has *already* taken place: the behaviour of modern human beings dealing with street traffic may indeed be disciplined, however it doesn't rest on *self*-constraint, at least not in the moral sense, but on *technical* constraints and technified methods of regulating traffic. The penalties for deviating from regulations are again external and are factored into one's expectations as such: deviations from appropriate matter-of-fact behaviour lead to accidents.

The external preconditions of everyday life, transformed over time into technical preconditions, have such a powerful effect on behaviour that individuals can progressively feel themselves absolved of ethical constraints. In tandem with a growing luxury economy, this leads to a waning of the puritan ethic and a substitution of technical norms for moral norms. Cases in point are the replacement of traditional handicraft ethics and business ethics by quality control, obligatory declarations and government supervision. A further case in point is the obsolescence of virtues such as thrift on the part of consumers. The standard example of technical norms taking the place of their ethical forerunners is the redundancy of traditional sexual morality as a result of the industrialized provision of effective means of birth control.[8]

The pervasiveness of technology

In the course of recent history, technology has come to have a civilization-defining significance, not only in Weber's sense (viz. as a system of rules geared towards efficiency), but as a system of material means. In everything from our society's technical infrastructure (e.g. railways and more recently the internet) through to the pharmacological means of regulating the way that we function physically and mentally, it is the technical conditions of life that determine how life is lived. In the process, the metaphor of the iron cage is now taking on a different hue of ambivalence. Taken at face value, the metaphor of the cage implies that technology is something external to our actions and social relations. Yet the technology that governs human life today is no longer external. On the contrary, modern technology has penetrated deep into social activity; technological systems are bringing about the technical regulation of human life in the very depths of our organic being. In the contemporary world, technology in short is not a cage in which human life can carry on protected and unchanged; it has instead become something like the skeleton of human life or, abstractly put, a sort of infrastructure of life. The all too human side of existence has thereby been reduced to a phenotype: what appears to wear all the bright colours of human individuality is little more than the game permitted by technical preconditions and their associated structures. In principle everyone is in the same position. Individuality becomes a surface phenomenon.

This situation demands a completely new conception of technology. Far from being understood as a set of rules to increase the efficiency of human action, technology was traditionally thought of as a system of material objects that we made instrumental use of to attain our ends. Human action and the ends themselves were taken as given.

That of course does not mean that technology, considered as mere means, was of secondary anthropological importance. On the contrary, since the time of the earliest philosophical anthropology – attributed to the sophist Protagoras – human beings, were seen as animals that have a primary need for technology because they would be unfit to survive on the basis of natural endowments alone. In Plato's dialogue, *Protagoras*, the eponymous sophist relates the myth of Epimetheus' creation of the human race. Epimetheus, so the story goes, had overtaxed himself creating other living beings, so this new creature, the human being, turned out to be a botched job: it was practically incapable of life. Epimetheus' brother, Prometheus, came to his aid by making human beings a gift of fire and hence all the handicrafts, or, as we would put it, a gift of technology. Ultimately, the god endowed these late-born creatures with the art of politics too, thereby enabling them to survive. Politics here is the last in a series of technical acquisitions that make good human beings' manifest natural deficiencies. (Nor is this rough view of the situation confined to the ancient world. In his anthropology of 1940, Arnold Gehlen[9] extends the myth, bolstering it with ideas derived from modern embryology. Gehlen speaks of humans as defective beings, as physiologically premature. Yet as necessary as technology is for humans, on Gehlen's conceptual picture it remains external, an extension or reinforcement of organic endowments, at most a substitute for organic endowments. Similarly, in 1877, Ernst Kapp,[10] the first philosopher of technology proper, interpreted technological structures as externalizations of organic structures and functions. Thus, for example, Kapp thought of the construction of bridges as a sort of recurrence of the lamella structure of the human femur and transport infrastructure quasi-literally as the nervous system of society.)

Today all of these utilitarian interpretations of the nature and the meaning of technology are clearly inadequate, if not downright misleading. Technical forms of organization do not make human action more efficient, they simply transform it. Mobile phones have brought about profound changes to social relations, above all among young people: fixed appointments are practically a thing of the past in a world in which it is always possible to make new arrangements at the last minute. The internet has redefined the sense in which society coheres *as a society*: we belong to society in the form of an email address; a homepage is increasingly the definitional mode of social self-presentation. And biotechnology has displaced the genetic boundaries between species – a trend that human beings could be caught up in as well in the long run.

Likewise, we can no longer comprehend the social effects of contemporary technology in terms of causal relations, as Marx originally attempted to do. In Marx, the development of the forces of production determines the *relations* of production; the productive forces (e.g. the machinery in use in a factory) and the relations of production (i.e. the relation between entrepreneurs and workers) – two different types of parameters – stand in a relationship of causal dependence: industrial methods of production *give rise to* an opposition between capital and labour. Invasive forms of technology, on the other hand, have *structural* rather than causal effects, a type of influence best understood using Foucault's concept of a *dispositif* – a conditioning factor that makes something else possible but also limits it, thereby giving shape to what it makes possible. Today we can speak of technology as a social *dispositif*,[11] a communicative *dispositif* and, indeed, a perceptual *dispositif*. What this means is that:

1 Our society's existing technical infrastructure determines what is socially possible today. Historically, this occurred with the introduction of technified means of transportation such as the physical movement made possible by railway networks. Today, however, it also means the movement of the personal data used to conduct a vast range of social business.

2 Today the possibilities of human communication are structurally shaped by communication technology. A typical example is the crowd dynamics of young people connected via mobile phones, gathering, dispersing and intermittently regathering in different places on the streets of cities at night.

3 The possibilities of perception itself are moulded by a vast range of technical media. Just as the categories of visual perception have long been moulded by photography (seeing something well has come to mean seeing it *in sharp focus*), what people are conditioned to *hear* has, in recent times, been increasingly moulded by acoustic technology as well.

In a technological civilisation, we might say, technology is no longer something that stands *over and against* human beings; it structures human life and social relations *from within*. Technical forms of organization, technical networks and equipment, the entire material aspect of technology, have penetrated into the depths of who we are, into our bodies and our communicative relations, in the process bringing about deep changes to the way our societies cohere. We are on the high road to defining ourselves in technical terms as human beings: to conceiving of the identity of a human being as a genetic fingerprint, to thinking of the biography of a human being as his or her electronic medical record and to thinking of learning as an interaction with databases and of society itself as an integrated network of personal computers.

Critical resources

Is this what we want? To be sure, there are people who greet the way that invasive technification is transforming human relations with euphoria. They have found their spokesman in Bruno Latour,[12] a figure who glorifies the increasing obsolescence of the distinction between nature and technology, and in Donna Haraway,[13] who emphatically greets cyborgs as a new form of human life. Yet is it possible that they too feel a lurking unease? The developments that they rush to praise were unintended; after all, in the past technology was understood as the amelioration of the conditions of human life, not as their transformation beyond recognition. The belief – which I have elsewhere called a *Baconian faith*[14] – was that technological progress was, by its very nature, human progress, *humane* progress indeed. Yet is there any denying that technological development is, in the final analysis, the result of economic and military competition? When such a development requires ideological justification, its advocates have recourse to the utilitarian argument that technology must be useful – naturally for human beings and society as they stand. The result of technological development though is that human life on the whole is not made better, but merely different.

Yet how is one to criticize the development of technology? Criticism always presupposes critical distinctions and a guiding idea of what rational conditions would look like, as Max Horkheimer might have said. Where today might the potential for making critical distinctions within our concept of technology lie and how might we bring into view the alternatives to our present way of dealing with technology? Technological utopias – what we find in science fiction – almost invariably present horror scenarios, unsettling visions that describe a path leading away from a world populated by human beings towards a world of cyborgs; William Gibson's novels, which have practically triggered a social movement of their own and a had real effect on actual technological development, are typical here.

What alternatives to the dominant trend of technological development have already emerged? 'Small is beautiful' was once a popular catchphrase that put centralized large-scale technology in the critical spotlight. There was the notion of *embedded technology* (Andrew Feenberg). And there were the twin notions of the *environmental* and *social compatibility of technology* (K. M. Meyer-Abich), which helped to make technology that was developed purely for the sake of economic efficiency into an object of criticism. To disabuse oneself of the nonsense surrounding so much technological development, it is also worthwhile trying to breathe new life into the old distinction between *useful* and *amusing* forms of technology that was formulated by the Renaissance technologist Samuel de Caus. And, of course, the age-old distinction between technology and nature is still an effective critical resource in some parts of the world, in practice if not always in theory. In the way that we eat, in the way that we relate to our own bodies, in insisting on face-to-face communication – this is where the potential lies, not just the potential to criticize invasive technology, but also to resist it. Practicalities are what it comes down to, for every purely theoretical or moral criticism is likely to be ineffectual if it is not supported by living habits and the unspoken givens of daily life; in a word, if it is not supported by *culture*.

Taking nature as a point of orientation and insisting on 'the natural' are still a valid basis for resisting the total technification of human life and relations. However, this is not the concept of nature pure and simple; nature in many parts of the world is actually also a *cultural* resource. Nature, we might say, is not an abstract concept, but an idea that is deeply bound up with our images of landscapes, with agricultural practice, with the work of innumerable conservation groups, bound up too with the way that human beings relate to one another and to themselves. Yet we would do well to ask whether these resources for coping with technology will remain at our disposal for long. Is invasive technification not precisely the sort of process that eats away at nature as a cultural resource? Our doubt arises because technology, in its invasive form, changes what our own nature is, what the natural world external to us is, and what our fellow creatures in the animal world are. It does so, moreover, *from the inside*. In our embarrassment, we fall back, at least reasonable people fall back, on prohibitions: we ban cloning, we stipulate that individuals' identity cannot be equated with their genetic make-up, we frame laws to protect embryos and laws to protect wildlife. And these are not bad things, though there are always lobby groups that would be glad to see such legal dam walls torn down.

The philosophy of technology

Paradigms

The question is why a philosophy of technology should exist in the first place, in any case something that is to be more than philosophy in the trivial sense in which one might speak of a philosophy of skiing or a philosophy of flower arrangement? Why should there also be a specific discipline, known as the philosophy of technology, taking its place alongside other philosophical disciplines such as ethics, aesthetics, logic, metaphysics and the philosophy of nature? What gives rise to these questions is the paradoxical situation that, on the one hand, the philosophy of technology exists in point of fact (viz. in the form of publications, curricula and professorial chairs), while as yet having produced little by way of a convincing intellectual paradigm. That may be because philosophy, in any case contemporary philosophy, belongs squarely among the humanities, and researchers in the humanities on the whole have rarely sought to see the phenomena of technology from the inside. Could this also be why their results have been paltry and, in some cases, downright dull?[15]

However, it must be said that this assessment is not quite accurate. Enumerating philosophy's subdisciplines already suggests a place for a philosophy of technology, namely as a counterpart to the philosophy of *nature*. When Aristotle, for instance, divides being in general into what exists of its own natural accord and what exists by virtue of technology in his *Physics*,[16] this indeed suggests the possibility of developing a philosophy of technology alongside the philosophy of nature; indeed, in light of the proliferation of technology in today's world, the need to do so is urgent. What mode of being could technical devices and objects be assimilated to? The question itself implies that the philosophy of technology would have to be an *ontology* of technological entities. With his notion of *Zeug* ('equipment'), Martin Heidegger's *Being and Time* might provide a pointer[17]: being of a technical nature is *Zeug*, with the existential character of being *ready-to-hand*. Technological objects, we might say more conventionally, are useful in a very particular sense; they fit inconspicuously into a field of meaning, only becoming noticeable if they cease to be useful, in which case they fall back into a different existential mode, that of being merely *present*-to-hand. Yet as illuminating as Heidegger's approach is, it really only suits traditional technology, e.g. handicraft technology; modern technology is another matter. Heidegger obviously saw the limitation and in his later philosophy went on to sketch a philosophy of technology that oversteps all limits.

Moreover, it is not true that the philosophy of technology is altogether lacking in paradigms. In addition to the ontological paradigm that derives essentially from Aristotle and Heidegger, we must at least make mention of the paradigms whose roots lie in anthropology and the philosophy of history. As we have seen, the anthropological paradigm goes back to the sophist Protagoras, or rather to Plato, who ascribes it to Protagoras in his dialogue of the same name. On this interpretation of technology, the guiding question becomes 'What is a human being?' Technology becomes a *humanum*, a human competence with the help of which human beings either compensate themselves for insufficient natural endowments or become capable of surviving in the first place. Technology on the anthropological interpretation thus belongs to the essence of being human. Human beings are by nature deficient beings, as Gehlen was to put it in the twentieth century. They are what they are, only insofar as they also develop culture. For Plato/Protagoras, social or political skill is part of the picture, not just physical equipment. In Plato, the capacity for politics also comes under the heading of technology; it is the *technē politike*.

The negative estimate of technology implicit in this concept (technology for Plato is merely a makeshift of necessity) already elicited a strong response in antiquity; in its modern reformulation in the work of Gehlen it is radically revalued;[18] for Gehlen, technology is no longer a sign of a limitation that animals are not subject to, but a mark of distinction, a capacity through which human beings surpass other animals. Using technology, human beings can be free of the given conditions of life, they can emancipate themselves from the world of nature. Through technology, humanity creates an environment that suits it; technology becomes second nature to it – an idea already to be found in the work of Marx. Admittedly, examples of the use of technology among animals were now identified too (dam building by beavers and the like); however, even if we assume that the use of technology is in the end something natural to animals, viz. a genetically inherited pattern of behaviour, the difference between the technological worlds of humans and animals nonetheless remains spectacular. Leaving inventors of the ancient world, like Hero, Archimedes and the fabled Daedelus to one side, since the advent of the modern world technological discovery has come to have a dynamism that connects it in an essential sense with ingenuity (Lat. *ingenium*). The paradigmatic technologist on this conception is not actually the craftsman but the *engineer*. In modernity, technology is no longer simply a means of survival, but a means of substantially enhancing life.

This is already to flag the transition to our next paradigm, which derives from the philosophy of history. The anthropological paradigm might be said to take for granted an unchanging relation between human beings and technology; it fails to allow for historical change, even on the technological side of the ledger. Yet the history of technology is a sort of history unto itself, and in relating it we can identify the major epochs in the relationship between human beings and technology. The most notable work to do so is not actually a work of philosophy, but a cultural history of nature, Serge Moscovici's *The Human History of Nature*.[19] Though one might not guess from the title, the book is also a cultural history of technology in the sense that it sets out the epochal differences in the characteristic relationship between forms of labour and concepts of nature. *Handicraft work* here corresponds to nature conceived according to the notions of matter and form, *engineering* is aligned with nature considered as an interconnected field of force and *cybernetics* with nature considered synthetically as a construct. We can also talk here of the epoch of one or another *dominant* technology, thus of epochs dominated successively by form-giving handicraft technology, by mechanics and industrial technology as organizing forces and by the controlling and ordering functions of information technology.

An interpretation of technology from the standpoint of the philosophy of history is actually to be found in the first philosophy of technology to explicitly identify itself as such. Ernst Kapp's philosophy of technology[20] could also be subsumed under the anthropological paradigm, viz. insofar as Kapp seeks to interpret various technologies as projections of human bodily organs. In doing so, he amplifies the Aristotelian notion of technology as a form of *mimesis* (imitation) by assuming that in technology human beings give external form to the structure of the human organism itself. Yet while the load-bearing structures of bridges may indeed resemble the lamella structure of the human femur and while the telephone network can certainly be interpreted as a sort of nervous system, it is of course easy to point to technologies that have no counterpart in the human body (the wheel for one). To be sure, there are grounds for interpreting technology as an imitation of nature. Nature's technical solutions prove time and again to be exemplary in terms of simplicity and efficiency: consider the stability of blades of grass from the structural point of view, the body shape of the dolphin from the point of view of fluid mechanics, the capacity of the human brain to store and process information. Yet imitation is anything but unconscious, as Kapp maintained. Today especially, it is something based on research, viz. in the field of bionics.

The decisive point of Kapp's philosophy of technology, however, lies in his integration of technological developments into the history of human consciousness. His *magnum opus*, *Principles of a Philosophy of Technology*, is in effect the materialist counterpart to Hegel's *Phenomenology of Spirit*. Just as Hegel tracks the way that spirit comes to consciousness of itself in the objective forms of its own works, so Kapp describes the way that humans become conscious of themselves as embodied beings in their technological creations. The conceptual figure in Hegel and Kapp is one and the same: self-consciousness is attained by making the implicit explicit. According to Kapp, 'self-consciousness thus proves to be the result of a process whereby knowledge of something external is identified as a knowledge of something internal'.[21]

Just as Hegel thinks that in his philosophy the world of spirit finally comes to consciousness of itself, so Kapp thinks human beings achieve true self-consciousness in his *Principles of a Philosophy of Technology*. In Kapp's quasi-prophetic world-historical terms, 'the conscious creation of technology might shine brightly in the foreground of human achievement; it remains nothing more than a reflection of the depths of the unconscious, the elaboration of a consciousness that was first redeemed by rudimentary tools'.[22]

While Kapp's philosophy of technology is hardly known outside specialist circles, Martin Heidegger's later philosophy of technology has had considerable influence in the history of ideas. What makes this all the more astonishing is that Heidegger consciously positioned himself at a considerable distance from all matters technological, both in life and thought. Yet it was probably for just this reason that he hit a raw nerve among those of his contemporaries who were disquieted by technological developments. His pamphlet-sized *Die Frage nach der Technik* (*The Question Concerning Technology*[23]) can be read as a radical criticism of modern technology, which in Heidegger's picture is a technology that reduces nature to a mere resource, destroys traditional forms of life (and thereby traditional forms of technology too) and in the end turns human beings themselves into mere raw material. Heidegger paints into his highly stylized picture of *Geschick* ('destining') everything that for the individual appears overwhelming about modern technological developments, everything indeed that might be soberly described in terms of the emergence of technology as a third cultural superstructure alongside the modern state and the modern market economy. Under the rubric of *Gestell* ('enframing'), technology itself becomes an epoch in the history of Being for Heidegger; it is the epoch when Being discloses itself through a technical *Herausforderung* ('challenging-forth'), and thereby withdraws all the more decisively. Yet in the process the role of human beings in maintaining the openness of Being, indeed in maintaining truth itself, becomes open to view – and it is here that Heidegger detects the possibility of deliverance from the epochal authority of machine power. It is impossible today to go along either with the subtly sanctimonious formula that Heidegger borrows from Hölderlin:

Wo aber Gefahr ist, wächst
Das Rettende auch.
('Where the danger is, there too//The saving power grows')[24]

or his associative style of presentation, haunted as it is by obscure readings of Greek and German etymology. Yet something significant remains of Heidegger's philosophy of technology, a view of technology that, like the entire ontological paradigm, goes back to Aristotle. To see what it is, however, we need to bring a fourth epistemological paradigm into the picture.

The ultimate source of the epistemological model is in Aristotle's *Nichmochean Ethics*, where he enumerates a series of types of knowledge or forms of *aletheia*. Reason, wisdom, science and ethical insight are on the list, but so too is technology.[25] The high regard for technology as an independent form of knowledge is all the more impressive in that technology does not even appear on the graded scale of types of knowledge that Plato sets out in the so-called 'analogy of the divided line' in his *Republic*. (Though one should not draw the conclusion that Plato's attitude to technology is therefore contemptuous; for example, that it involves looking down on manual labour as too banausic. Above all, it was the expertise of craftsmen that best met Socrates' challenge to his contemporaries' claims to knowledge. Plato's thoroughgoing rationalization of all forms of public activity calls in a similar vein for the development of a specific technique, a *technike*, appropriate to each activity.[26])

In Aristotle, technology is characterized as a form of knowledge that guides the fabrication of things. Making things (*poiesis*) for Aristotle is a form of human action to be distinguished from *praxis* – the (self-contained) fulfilment of life's activities. Making has its aim in a product or work, whereas the goal of *praxis* lies in itself.[27] This is the late Heideggerian point of departure; for Heidegger thinks that with the advent of modern science all forms of knowledge have come under the domination of the technical model of knowledge. Since Galileo, indeed, it has been in technical contexts, through experimentation and measurement, that nature has been an object of knowledge; according to Descartes' maxim, we should seek to understand nature as the expert craftsman understands the products of his craft: to know something means knowing how it can be fabricated. Science thus becomes an exercise in technically reproducing nature: that, at least, would be the identifiable meaning of Heidegger's speculations about the history of Being. The epistemological model of technology implicit in the technologist's view of the world becomes the dominant form of knowledge in modernity. The critical potential of Heidegger's philosophy of technology is hence essentially as a critique of the modern world view – a critique of science similar to what we find in Goethe's *Theory of Colour* and many strands of twentieth-century phenomenology. Goethe and the phenomenologists object to the technical model of knowledge by opposing it to a knowledge that places a premium on ethical orientation and individual self-cultivation (*Bildung*) as well as systematic description. Critical theory approaches the same epistemological circle of thought from a different direction, objecting to the technical model of knowledge on the grounds that it is technocratic. What Aristotle calls *poiesis* and *praxis* has its place in a critical theorist such as Habermas under the rubric of purposive-rational action (or interaction). In opposition to technocratic knowledge, critical theory posits a form of knowledge capable of effecting practical change through critical reflection about existing social circumstances, reflection that is guided through and through by what Horkheimer called 'an interest in rational conditions'.[28]

Once these four paradigms of the philosophy of technology have passed muster, we might well ask why, in spite of such splendid beginnings, we still have the impression that a true philosophy of technology hardly exists. Our disquiet here can only stem from the fact that these philosophies don't really help us to deal with the problems that technology actually poses. Speaking generally, we can say that they all view technology as a domain unto itself, something that human beings may indeed stand in essential relation to, but that in the final analysis remains foreign to them. Human beings are what they are; they cannot but make use of – or, as Heidegger says, cannot but be made use of by – technology. What will concern us in the present book, however, is invasive technology: technology with the potential to alter what we understand as the specifically human aspect of human life.

The main consequence of approaching technology as a domain unto itself is that we set out to grasp something like the essence of technology. This is understandable in a philosophical context and of course goes back to Socrates' revolutionary fascination with questions of definition (the question *ti estin* – what kind of thing is that?). However, the decisive question today cannot be the question of the essence of technology, but rather the problem of technology as a *process* – as technification.

Technology considered as technification is not something external that supplements what is essentially human, but something whose role in shaping *what it is to be human* needs to be the basic theme of our discussion. The concept of technification requires us to comprehend technology as part of a dynamic field of historical tension which can generate resistances and, in certain circumstances, release the potential for criticism.

In specific terms, we can say the following: in making an in-principle distinction between technology and nature, *the ontological model* of the philosophy of technology contains the theoretical seeds of a critical approach that has been and will likely continue to be effective in Europe in particular.[29] The potential difficulty here is that the technification of nature could mean that the concept of nature loses its normative and hence its critical relevance vis-à-vis technology.

What speaks for *the anthropological approach* is that it takes technology's meaning for human beings seriously, in particular its meaning for human beings' conception of themselves. However, this approach revolves too narrowly around an instrumental model of technology; it views technology as a means to an end. It is not that the anthropological paradigm has somehow become obsolete because it has been overtaken by technological developments themselves, though it is true that modern technologies can no longer be understood using the concept of utility or from the point of view of instrumental relations. Rather, it is precisely in an anthropological sense that this strand of the philosophy of technology is outdated. Historical anthropology[30] has taught us that we can no longer speak of an invariant human essence. Crucially, we have today reached an historical point where we no longer conceive of humankind by differentiating the human from the animal, because, on the one hand, our proximity to the animal kingdom is something we no longer shrink from, and because, on the other hand, the gulf between the human and the animal, in part precisely due to the development of technology, has become both unmistakable and unbridgeable.

Most of the approaches to the philosophy of technology with roots in the *philosophy of history* come across nowadays as grand but fantastic speculations. If they fail to speak to us, this is precisely because they tend to view technological development as an unconscious process or as an overpowering fate. They function like metaphysical glosses on the old trope of technology as a demonic power. In the background of Heidegger's later work on technology, one detects a radical critique of modern technology. But precisely in stylizing technology as a 'destining' that human beings cannot but conform to, late Heidegger again leaves us at the mercy of historical developments.

The *epistemological paradigm* of the philosophy of technology provides the most promising basis for critique. Yet alternative forms of knowledge, such as phenomenology, which are well suited to articulating what distinguishes our specifically technological experience of the world, are yet to be used to critically describe the processes of technification. Instead, the alternatives (Newtonian optics versus Goethe's *Theory of Colour*, the world of science versus the everyday lifeworld) have simply been placed alongside one another without any reference to the dramatic changes that arise when one point of view begins to be displaced by its radically divergent counterpart. In particular, what is overlooked on the epistemological model is that *things themselves* evince different structures when they come within the purview of one or another type of knowledge. In short, what is lacking here is an awareness that a technification of our ways of seeing also alters the very world in which we live.

Philosophy of technology as critique

What would a philosophy of technology that sets out to understand the meaning of technology in light of the processes of technification look like? For example, do theories of rationalization or modernization already provide such a philosophy? The latter indeed are theories that attempt to pinpoint the defining features of ongoing change – both in the lives of individual human beings and in human society at large. Examples are to be found in the works of Max Weber and Jacques Ellul,[31] and although Weber and Ellul's theories are actually social theories, they could also be regarded as philosophies of technology were one to accept their preferred definitions of technology/technique as something like a system of rules oriented towards efficiency. As we have noted, one can speak in this sense of a pianist's technique and of techniques for producing goods. As we have also seen, such a concept of technology/technique has its origins in Aristotle's notion of *techne*, though the emphasis in Weber and Ellul, in typical modern fashion, is not so much on *what* is fabricated, but on the efficiency of the fabrication process itself. No doubt, this procedural, efficiency-oriented notion of technology is very well tailored to capturing the dynamics of processes of technification and the totalizing tendency associated with those dynamics. In the modern world, technology is indeed no longer an object or a set of objects that we can neatly separate from others; rather, it is something omnipresent, something that pervades everything from the life of the individual to the way we shape the environment as a species. The only problem is that this concept of technology is in a certain sense too idealistic. It completely neglects the concrete forms that technology assumes, from craft tools and machine tools all the way up to the infrastructure of our communication and transport systems. Granted, these systematic arrays of material instruments also serve the purpose of rationalization and increased efficiency (in such systematic arrays, the principle of rational efficiency takes on a material form as it were), yet they also condition human relationships in quite a different way to mere rationalization; through their very existence, they set the conditions of possibility of the lives of individuals and societies. Driving a car is not simply a more efficient form of walking, to phone someone is not simply to speak to him at a distance, a sleeping pill is not simply a faster means of falling asleep, and the integration of our society via the internet is no mere rationalization of human interaction. As a systematic array of material means, technology does not simply leave the human relationships whose fulfilment it serves as they were, it transforms them structurally. And it is these structural changes which exceed the scope of concepts like rationalization and which ought to be referred to as technification in the true sense.

We have to regard technology as a *system of material means* because the interpretative model provided by hand tools no longer suffices in light of modern technological developments. The meaning of modern technology can no longer be grasped by considering isolated technical devices because *what the latter are* can be understood today only in terms of their integration within systems. In general, technical devices no longer perform the tasks that they are supposed to perform when decoupled from the systems that they are usually embedded in. Thus, a car is no longer functional in the absence of the relevant material infrastructure made up of streets, petrol stations and traffic signals, as well as more abstract, socially defined structures like road rules, insurance and the like. Indeed, not even something apparently as simple and self-contained as a wristwatch functions as it should without reference to global measurements of time.

Technology that takes the form of systems of material means also requires us to disregard the concept of functionality in certain respects. Though Heidegger is right to suggest that the classical technological object, the hand tool, has to be understood in terms of its usefulness, this is not something that can be said of modern technologies in general. Initially, the latter may well be devised for particular purposes; however once in existence they can serve other purposes as well. A classic example is the internet. Originally developed as a tool for military communication, today the internet fulfils a vast array of functions – including the role of a new public realm, a virtual marketplace and a social network.

The basis of this multifunctionality is precisely the materiality of these new forms of technology. Once installed, they function independently of their creators' intentions. New users can devise entirely new usages. We have reached the point where it is possible to speak of the technological legacy of a society (as opposed to that of individual inventors and innovators) – even of the technological legacy of humanity as a whole. Technological infrastructure outlives the generation of its creators and, in turn, supplies the basic preconditions for the lives of subsequent generations – precisely the sort of effect that one can call technification in an emphatic sense of the word. Clearly, in a certain sense, the whole realm of culture does this too; like technological infrastructure, culture is a sort of historical legacy that marks out what is possible in life from one generation to the next. Yet it does so only insofar as it is continually assimilated anew through processes of socialization and learning. Technological infrastructure, on the other hand, supplies a set of quasi-natural boundary conditions; Marx speaks quite rightly in this regard about a second human nature. To be sure, each generation defines and redefines technological possibilities. Each assimilates the given technological preconditions in its own way. Yet to begin with socialization requires *accommodation*; it is above all up-and-coming generations that take pre-existing technology as a kind of natural given.

Under the conditions of contemporary life, technology is thus not simply a means that we already possess or that we have to invent in order to reach certain ends. Rather, it is primarily *what we find given as a precondition* – a precondition that defines in all cases what human life can be. Technology has become a sort of infrastructure of human life itself, a medium of human life.[32] What it is to travel in our contemporary lifeworld, what communication is, what work is, what perception is, can no longer be determined independently of technical structures. And yet we should avoid focusing solely on the limitations of our predicament here, for technology is clearly what also makes many previously unthinkable forms of human existence possible in the first place. A classic example is the enormous expansion of the human power of sight by means of the telescope and the microscope. Travel is another obvious example. Extending the range of the humanly possible was, indeed, precisely what stood in the foreground of modern scientific and technological development from the outset, for example in the work of Bacon.[33]

If, today, we are compelled to emphasize the limitations associated with technology and its intrusion into so many aspects of human life, it is important to bear in mind that limitations do not necessarily imply the opposite of an enlargement of the field of the possible, but can also mean limits in the sense of formal conditions: by penetrating into the way that human beings behave and relate, technology defines what behaviour and relationships actually are. To stick with examples for a moment: compared to seeing something with the naked eye, seeing something through a telescope is a different type of vision; compared to train travel, air travel is likewise a different type of travel, involving the substitution of an experience of a physical path between two places with a period of time spent in the hermetically sealed confines of a plane's cabin. And clearly falling asleep with the help of sleeping pills is something structurally different to sleep that involves losing oneself in a gradually dissipating haze of mental images. For the purposes of our proposed investigation, to capture both sides of technology – its limiting and its enabling aspect – we take up a concept Michel Foucault reintroduced into philosophy: the *dispositif*.[34]

In the context of a philosophy of technology whose main theme is the processes of technification, it makes sense to call technology a material *dispositif*. Technology is thus defined relationally, for the concept of a *dispositif* contains one open variable. In individual cases, we will thus speak of technology as a *dispositif* of communication, as a *dispositif* of perception, etc. Our proposed concept of technology defines technology as the condition of the possibility, the guiding principle of the arrangement, of human behaviour and interaction insofar as behaviour and interaction are determined by *material* arrangements rather than discipline, education, strategies of individual self-cultivation or social convention. The notion of 'the material' should not be taken in too narrow a sense here, for clearly such arrangements are just as likely to consist of software as they are of hardware. A decisive point of our conception of technology is that it is independent of the individual human beings who might or might not have accommodated themselves to a particular set of technological arrangements by acquiring this or that technical skill.

A philosophy of technology that deals with processes of technification is, by its very nature, critical insofar as it sees technology as standing in a relation of tension to what it is a technification of. The description of individual processes of technification as we conceive it is already critical in the Kantian sense, viz. insofar as such description brings to light what the processes of technification exclude in the very act of enabling particular forms of human relationship and behaviour. The notion of a *dispositif* already expresses this very ambivalence.

The works of Hubert Dreyfus and Joseph Weizenbaum each contain critical philosophies of technology in this sense.[35] Both can be located on the margins of a critical theory of technology.[36] The task of such a critical theory – which is one of the tasks that the present work sets itself – would not merely be to bring the processes of technification into critical confrontation with what they are technifications of, but to critically assess technification in light of *alternative* technological possibilities.

2

Science, technology, civilization

Civilization in the age of technoscience

In this chapter, we will be dealing with the meaning of science and technology in our lives, both personally and socially. My aim, however, is not to pose this question of meaning primarily as a scientist or a philosopher, but in a form that anyone could pose it insofar as he is affected by technology – affected, one could say, *by the very fact of its existence*. Let me start with some examples: it seems true to say that our way of communicating with each other has been changed, not just by the use, but by the very existence of the telephone and a host of other communication technologies; that the organization of our social life has been changed by social theories and by the existence of computers; that the way we feed ourselves has changed with the rise of chemistry and scientific agriculture and the systematic study of nutrition; and that the way we relate to our bodies has changed as a result of the existence of pharmaceuticals. With these examples in mind, we could say that the most urgent task of reflection is to establish what it means to live in a world that has been shaped fundamentally by science and technology. However, that way of framing the issue immediately raises a further question: are science and technology the same thing? Let me try to dispose of the latter question as quickly as possible, for it has already given rise to a vast literature.[1]

Rivalries and status wars between scientists and engineers are essential to understanding the answers that have been given. It has been argued variously that scientists produce papers while technologists produce things; that technology's role is to apply scientific knowledge; that science aims to be true, while technology aims to be useful; that science deals with the world of nature, while technology deals with a world of man-made objects. However much truth there may be in these kinds of general claims, to me it seems clear that science and technology represent two sides of a single human project and that they cohere in important ways: the differences between them are differences of accent or orientation. Technology, for instance, is by no means the mere application of scientific laws, but a process of modelling nature according to social functions and demands[2] (though it must be said that such modelling already occurs in science itself). When people argue that scientific knowledge is the basis of all technological endeavour, they could just as easily put it the other way round and say that technology is the basis of science. However one sees this issue, science and technology become relevant in the personal and social sense *when considered together*. What we might call the *scientification of our lifeworld* does not mean that we adopt a scientific attitude to our lives, or that our everyday knowledge of the world is transformed into scientific knowledge; it means, on the one hand, that technological objects and technological products play an ever greater role in our lives, and, on the other hand, that large swathes of practical activity are delegated to technical experts. This is the reason why science and technology form a single indistinguishable whole from the point of view of the layman.

Nor is the lay view without a basis in reality, in this case historical reality: for there is indeed a unity to modern science and technology that follows from the fact that they have a common source in the age of the Renaissance. It was in the sixteenth century that mechanics found acceptance as a science, and nature was studied for the first time in history under systematic technological conditions (i.e. *experimentally*) using technological models like the mechanical clock. Yet the times were not quite ripe for the emergence of a truly technological science; the unity of science and technology was only to come into practical effect in the nineteenth century.[3]

Since then, the points of interconnection have become much more numerous and complex. (Nonetheless, many writers today still insist on the difference between science and technology in deference to Karl Popper's notion of the hypothetical nature of scientific knowledge. If science is fundamentally hypothetical, then what engineers know must be different to what scientists know – or so the thought goes: engineers can do little with mere conjectures. Fortunately for everyone though, the majority of scientific knowledge is by no means just a matter of probable hypotheses.)

When we look at how things stand 400 years after the birth of modern science, it seems far more accurate to put science and technology on one side of the equation and a general sphere of technics, including individual technical devices and large-scale technical systems, on the other side. Of course, one can think of technical devices in Marx's terms as 'frozen knowledge'; a machine is indeed a kind of object in which scientific theories and concepts take on a material form. (Interestingly, many of the basic structures of our society are, in large part, a 'frozen social or political science' in the sense that they are based on social relations that have been reconstructed on the basis of scientific knowledge; and indeed this, in a sense, is precisely what our society's fundamental drift towards rationalization and scientification amounts to – the reconstruction of both our natural *and* our social environment in accordance with the deliverances of scientific wisdom.) Yet there remains a difference between the *representation* of knowledge in the form of scientific theories and the *incorporation* of knowledge in machines and other structures of material reality; it is only in the first case that it is clearly true to say that knowledge is represented *as knowledge*. The really important point is this: for the individual human being who is affected by scientific and technological change, there is no difference to speak of here – for the layman, the natural and the social world might have become more perspicuous or easier to control in some senses; however, both the content of scientific papers and the underlying structure of a world reconstructed in the image of scientific knowledge remain a matter for specialists. (The same, incidentally, might be said from the point of view of information theory, which views scientific theories and material structures as simply different forms of the same knowledge.)

To return to our initial question – *what science is* can only be settled by looking, among other things, at its many and striking effects on human life. Science has had such a profound influence on human life that it can no longer be adequately described just as a type of knowledge or as the product of a certain social group (scientists and the scientific community), or indeed as a way of relating to objects generally. Science must be seen as an epoch-defining phenomenon – a cultural form that has shaped a certain period of human history and will in future likely shape the very course of natural evolution.

Since the nineteenth century, there has of course been no shortage of theories that attempt to come to grips with science in epochal terms. They include the romantic interpretation of science central to German idealism, which tells us that scientific and philosophical development is an evolutionary process in which nature arrives at self-consciousness. They include Comte's theory of cultural evolution, which tells us that the age of science – the age of positive history – is what follows the theological and metaphysical ages of human history. They also include Weber's theory of rationalization, which describes a long-term developmental process which encompasses the whole of European history, and which in the twentieth century comes to incorporate the entire world. Lastly, they include all sorts of theories that attempt to grapple with the meaning of science and technology for human life in recent decades: the Marxist theory of technoscientific revolution, Bell's theory of post-industrial society, and a variety of attempts to show that technically educated intellectuals form a major new class of their own.[4] My intention here is not to give an assessment of individual theories. Nonetheless, they provide the backdrop for the argument of the present essay. Instead of analysing existing theories, my aim will be to capture the effect of technology on human life in as direct a way as possible and in two key dimensions, viz. anthropologically and sociologically. In other words, my prime concern will be a question about the *meaning* of science – the meaning of the scientification of human life – for individual human beings and for society as a whole.

Technological civilization is as yet only a partially realized endeavour, something that we are in the process of producing. I use the word 'produce' advisedly here – even though the conditions of human existence in our technologically advanced world are not the realization of any set of straightforward technical plans or intentions, but in many ways the side effects of a process of scientification we pursue for quite different reasons. Life is something that we want to make simpler, work is something that we want to make more effective and communication is something that we want to make faster and easier. Our overall goal is to improve human life, but in doing so, we are altering its very structure. Human life in the age of technological civilization cannot be deemed better than life as it was lived in the past simply because it is different. The differences between past and present are regularly tallied up as losses on the balance sheets of cultural critics. And it is often difficult to avoid their language and some of their tone, even when one sets out to talk neutrally of changes, instead of benefits and losses.

Human self-relations

Let me begin by looking at one change that technology brings about something that many people find particularly dramatic because of the way it touches on a fundamental point of human self-understanding. I am thinking of the way technology makes ethics redundant. Social scientists tell us that the function of morality once lay in the realization of particular social goals through the regulation of individual behaviour. The effect depended, of course, on the fact that individuals themselves were unaware of the social purpose of the moral norms they obeyed; from the individual's point of view, morality's dictates were solely a matter of what was *morally* good or *morally* required. When, at the start of the nineteenth century, Malthus expressly appealed to individual morality in the hope of halting the runaway growth of the population, his appeal alone was an indication that an era of history had come to an end. Today, the Malthusian social goal, a lower birth rate, is something that can be reached directly by technological means – in this case, contraceptive technology. Yet the complex consequence of that simple fact is that the greater part of sexual morality has become superfluous.[5] My claim is *not* that life within a technological civilization will, in future, be regulated technically rather than morally, let alone that it *should* be regulated in a exclusively technical sense; the end of morality is certainly not in sight. My point is that as a result of technological development, morality in the future will be a *different kind of morality*. Stripped of its social function, morality will become a kind of luxury. One foresees ethics moving closer to aesthetics (moving back, incidentally, towards the aesthetically modulated ethos of the aristocratic societies of the early Greek world).

The direct path that modern science and technology enables us to take to particular social goals, which were once reached by indirect ethical routes, goes to the heart of what modern science and technology *are*. The relationship between contraception and sexual morality is hardly an isolated one. Something similar is at work in the way modern human beings, both men and women, relate to their own bodies and souls. The typical attitude that science and technology – or, more exactly, the profusion of sophisticated technological products – gives rise to means that we are increasingly inclined to boost immunization rates rather than improve general living conditions in order to prevent tuberculosis; that we reach for sleeping pills instead of counting sheep in order to get to sleep at night; that we turn on a television instead of heading out into the world in search of new experiences. The knowledge science provides of the actual constitution of physical reality and the means that technology provides for reconstructing physical reality make the direct path to the satisfaction of our desires irresistibly tempting.

Let us look a little more closely at the attitude human beings characteristically adopt towards matters of body and soul in a technological civilization. Clearly, scientific knowledge of the human body is a form of *external* knowledge. For science, the decisive experience of the body is not *self-experience*, but the experience of *the other* – a technically mediated experience of the body that is, in no essential sense, the individual's own. Anatomy and physiology are the foundational disciplines of modern medicine. The scientification of our way of relating to our own bodies means adopting an attitude towards ourselves as embodied beings that is reminiscent of the relationship between doctor and patient. In short, we often treat our bodies as if they were not truly our own – we treat them as if we were able to see them entirely from the outside.[6] At this point, it becomes clear that the application of scientific knowledge, in this case the technical knowledge of contemporary medical science, does not simply mean that we act on knowledge instead of acting in ignorance. It means rather that our very way of relating to ourselves has changed.

When we turn to our way of relating to ourselves as *souls* (i.e. as conscious beings with a depth of inner experience), the position that we are in is not quite so easy to read. Although psychotropic drugs are often produced with the help of psychological and sociological research, those who consume them hardly say to themselves 'I have such and such a psychological need and I should satisfy it by such and such means'. Why not? The most likely explanation lies in the fact that human needs are amenable to such a wide range of interpretations. Putting it simply: I can say, for example, that I'm going to see a film because I like movies and *not* because seeing a film will fulfil certain functions within my overall psychic economy. And yet seeing the film does fulfil a function in a psychic economy. To get to the bottom of what this means, we need to go on a slight digression – though one that should be useful, for it will bring out another defining feature of technological civilization.

To live in a technological civilization means living in an environment that demands of the individual a cool, objective, decidedly *unemotional* stance towards reality. It also means that the world around us is rigorously organized with a view to guaranteeing security – ideally in such a way that nothing dangerous or damaging can befall the individual. Now, the upshot of those two characteristic traits of our social world is that a large part of the individual's emotional capacity lies fallow. On the whole, it has to be said that within a technological civilization, emotional life usually takes place in a detached part of reality, a second world populated by fictions. The rise of contemporary technological civilization is bound up with the development of an enormous imaginary domain generated continuously by the mass media, whose prime function lies in satisfying emotional needs. (Not that the process is without its preparatory historical phases – the decoupling of human emotional life from reality and the cultivation of fictional worlds in fact already begins in the early modern period and spreads progressively throughout all classes of society, starting at the top, often going hand in hand with the development of a specifically objective attitude to the world; one might think here of the extensive consumption of novels by the bourgeoisie of the eighteenth century or the enthusiastic moviegoing of the twentieth-century working classes.)

A further anthropological shift brought on by technological civilization consists in the externalization of social constraint – a development whose far-reaching effects we can gauge when we look at the broad sweep of twentieth-century social history from the point of view of what Norbert Elias dubbed the civilizing process.[7] A technological civilization, in certain respects, represents a reversal of the course of events that Elias thinks marked the advance of civilization; one of the latter's most important aspects is the *internalization* of constraints – as civilization progressed, Elias demonstrates, what had once been a response to *external* compulsions, viz. peaceful predictable human behaviour, became a by-product of *internal* mechanisms – a kind of side effect of compulsions that the individual exercised over *himself* (conscience, the superego and the like). From a contemporary point of view, however, it is becoming quite clear that Elias obtained the results that he did – and in a sense was only able to conduct the research that allowed him to obtain those results – because he was living at a time when the wheel of history was turning. Sexual morality was becoming more permissive, human beings' general sense of shame was receding, table manners were becoming more liberal, and schooling was becoming gentler and less disciplinary. Today, in fact, we can see the deeper historical rationale behind these apparently so liberating developments: internal constraints were, in effect, becoming superfluous because their function was being taken over by the external world.

The fresh tranche of external constraints was no longer imposed by individuals or groups exercising power as the masters of early modern societies had once done; they had little to do with the application of any sort of transparently repressive force. What was involved could be called the *structural* exercise of force. The new constraints were constraints imposed by the technical mechanisms of social existence, by technological infrastructure. One example can stand for many. It is astonishing to realize just how much emphasis was placed on punctuality in education in early modern Europe at a time when punctuality was relatively unimportant as a matter of social practice. Today, by contrast, punctuality is of paramount social importance, and indeed time itself sometimes almost seems to have become a fantastic system of total social co-ordination. Yet educationalists no longer show any great interest in punctuality. The reason for that ought to be clear: in today's world, punctuality is no longer the hard-won fruit of systematic inner discipline, but an effect of external compulsion – a kind of by-product of the automatic action of the very technical conditions of daily life.

The relaxation of intrapsychic constraint is also related to a fundamental change in the conception of what it is to be a normal, psychologically healthy person within a technological civilization. One of the concepts that has come to play an absolutely crucial role in the social psychology of recent decades is the concept of personal identity.[8] Part of the reason is to be found in the increasing frequency of the phenomena of *diffused identity*. Coming at the problem from the opposite direction: what the emergence of the concept of personal identity points to is that in recent decades human beings have found it increasingly difficult to integrate the various parts of their lives into a single personality. In the long run, the imperative to form a well-defined single personality looks to be losing some of its force as the conditions of everyday life (work and communication above all) change in certain ways. In recent times, for instance, it has become clear that many people, maybe even the most successful, no longer feel the need to integrate the roles they assume in the workplace and the family, and indeed if work itself were no longer to take the form of a social role, i.e. if work were no longer understood as a particular type of expression of one's personal being, then one could imagine a situation in which the social need to construct a well-integrated personality might disappear altogether. If work were to lose its traditional psychological function and become nothing more than the exercise of particular skills or a strategically advisable way of spending a particular period of one's life, then the effect on human self-relations would be momentous. And, indeed, there are signs that social technification is carrying us in just this direction.

Social relations

The obsolescence of ethics, the instrumentalization of the body, the decoupling of emotional life from reality, the externalization of constraints and the rigorous separation of personal life into functional sectors – all are changes that affect what it means to be a human being in a technological civilization, and all affect the very structure of the human personality. From individuals' modes of *self*-relation, I would now like to turn to the basic characteristics of *social* relations in a technological civilization. This, in turn, should bring into view the specific social structures that are becoming evident in our increasingly technified age.

To start with we need to frame a concept of technology which has a solid basis in social science. Those working in the field have rightly emphasized that the task requires bringing 'things' back into the sociological domain.[9] Sociologists have long thought of technology as a type of capacity or skill, as a process or ensemble of processes. For Ellul, for example,[10] the concept of technology includes every methodical procedure we value in terms of efficiency. We must take the opposite tack and attempt to come to grips with technological objects in terms of their place within civilizing processes – otherwise, we risk interpreting the technification that has come over us as a straightforward continuation of a process that Weber identified long ago as a basic thread of Western history: the advance of a narrow *instrumental* conception of reason. For us, the social meaning of technological objects and products goes deeper than that. In the final analysis, automated production is *not* just a more efficient form of labour, the telephone not merely a medium for long-distance conversation. At the end of a 200-year long historical development, natural science was forced to concede that the thermometer was no longer simply a means of improving on our natural bodily sense of temperature because *what it was to have a natural bodily experience* had changed radically in response to the existence and use of the thermometer itself.[11] And so it is that in social science we have to come to terms with the fact that technified social structures too are not simply more efficient variants of their pretechnological equivalents – that for the most part, in being technified, social structures and social action simply become different rather than more efficient or definitively better.

The way human beings relate to the natural world can serve as a point of interpretative departure. We can define technology as a materially appropriated nature. By this, however, we do *not* simply mean that any use of the powers or materials of the natural world ought to count as technological; nature in that sense is something we already make use of with every breath we take. The material appropriation of nature on our reading presupposes the isolation and purification of natural materials; it implies a rearrangement of nature in accordance with specific social functions that involves a partial suppression of nature's spontaneous activity. (It is worth mentioning in passing that in the process nature's spontaneous activity becomes latent, to be manifested negatively in deviations – for example in rust.)

It is in this sense that technology might be called socially appropriated nature. Yet it would be a mistake to leave the analysis there. The relatively abstract concept of material appropriation can also be applied to society. Technology, we might say, is a kind of material self-appropriation of *society* as well as nature.[12] What does that mean though?

Let us think through the problem of technology as appropriated *nature* a little further before extending the analysis to society. Around Marx's time, during the first half of the nineteenth century, it became necessary to make a distinction between machines and hand tools, and more generally between technology and craft. In chapter 13 of *Das Kapital*, Marx produced a string of definitions of machine technology. The most famous of these definitions tells us that the machine is frozen natural science, and it underscores one aspect of technology that at the time was yet to have become decisive, though it was, in fact, one of the fundamental characteristics of modern technology – its increasingly scientific orientation. Of course, technology today is still a materially appropriated nature: it is based on the fabrication of pure substances, the quarantining of environmental influences and the specification of precise parameters. Yet on the one hand, this material appropriation of nature already takes its lead from a scientific appropriation of nature, and, on the other hand, the mastery of nature by means of technological devices is based on *explicit* knowledge; it does not involve working *with* nature but actively *regulating* it. The mastery of nature that modern technology makes possible hinges on the fact that we know exactly what physical laws are obeyed by the processes that are at work in individual devices. Though I say 'we', I do not, however, mean the individual human being actually using the individual device, since knowledge about the processes at work in the device is itself objectified; such knowledge can be built into the machine itself in the form of a control system, meaning that the human operator does not strictly control operations, but sets off the workings of an internal control mechanism. The contrast is clear: the use of hand tools and other forms of traditional technology required a sort of material appropriation of nature that became increasingly irrelevant in the transition to machine technology; it called for an adjustment to nature – a considerable degree of bodily intuition and skill – whereas modern technology calls first and foremost for scientific knowledge.

If the connection between material/practical and intellectual/theoretic (scientific) appropriation is indeed an intimate one, what does this mean in terms of the technification of social relations? The first point to recognize is that the intellectual self-appropriation of society has its mainstay in *social science* – the sort of science that emerged from the close relationship between the state and the generation of statistics (what was later conceived by Comte as a kind of physics of social relations). The goal of positivist social science is to make society amenable to control through the collection and analysis of data. Yet this presupposes a form of material appropriation – something quite analogous to the fabrication of pure materials, definition of boundary conditions and control over parameters we saw was paramount in the natural sciences. To make society more manageable through the use of knowledge, society itself has to be organized so as to conform to the demands of knowledge: social processes have to be functionally differentiated and structured according to conceptually well-formed models, social activity has to be disciplined enough to make it possible to generate meaningful data – disciplined in such a way that human beings' social roles and the social consequences of their action do not produce a mere mass of statistical irrelevancies.

It seems we have succeeded in developing a concept of technology that incorporates both nature and society, as well as stepping beyond the identification of technology with skills or processes. In short, we can now say we are dealing with technology whenever the social appropriation *either of nature or society itself* is at issue, whether the act of appropriation is essentially material or intellectual. And we can speak of specifically *modern* technology and specifically *modern* society in cases where knowledge of the way a machine *or of the way society* functions is itself objectified, e.g. where that knowledge becomes accessible in the form of a control mechanism. (Incidentally, a controversial exercise in social self-appropriation of this sort was the German Federal Republic's original plan for a national census in the early 1980s; strictly speaking, the result was to be no mere statistical aggregate, but a one-to-one model of the population for the purposes of computer simulation and precision management. Resistance on the part of numerous groups led to a constitutional challenge and the proposal was soon shelved – though, unfortunately, the precise copy of an entire society that was being proposed can also be compiled in the absence of census figures.)

What the concept of technology that has arisen within positivistic social science boils down to is a notion of society itself as a kind of machine – something resembling Lewis Mumford's concept of the megamachine, in this case not applied essayistically à la Mumford to the pyramid-building society of 3000 BC,[13] but with methodical rigour to the contemporary social world. Though it seems useful, given the rise of social science, to characterize modern society as a knowledge society,[14] it is important to stress that the knowledge in question is *not* knowledge of what society is *in itself*, but knowledge of society as it has been organized – systematically drilled we might say – with a view to the production of knowledge of a very particular kind.

Technological objects, technological structures

We now need to pursue a different train of thought. Our first aim has been to win back a notion of *things* for the social sciences – to develop a concept of technology that did more than merely update the theory of instrumentally rational action to take into account new technological developments, but instead brought technological objects into the picture. Our next task is to show why it is *technostructures* rather than *individual technological things* that need to take pride of theoretical place in social scientific analysis. I will use the term 'technostructure'[15] instead of speaking of technical or technological *systems*, first, to emphasize that the type of formation we are interested in can be the technological structure of *something else* (e.g. a technostructure that has been *imposed on* the natural world or society rather than an ensemble of technological objects conceived for technological purposes), and second, because the concept of a system is narrower than that of a structure: by a system we understand one particular type of structure with specifiable boundaries, a specifiable internal coherence, etc. By stressing that it is technostructures that are of prime importance for our society rather than individual technological things, we want to bring out two related points: first, that as a rule individual technological things only *are what they are* as part of broader networks, and second, that in today's world technology is no longer a mere means to an end, but a form of practice. Of course, it is not difficult to point to individual technological things that we can make use of in isolation from one another. A wristwatch seems to be an object of this latter sort. A telephone, on the other hand, obviously has no use whatsoever except as a point of connection within a wider communication network. On closer inspection, however, it becomes clear that many technical devices that appear designed for use in isolation also only perform their intended functions when they are connected to a network, or when they are used in the context of a broader structure: a car, for example, is really only serviceable as a car along with a network of streets and roads, a network of service stations, an insurance system, a system of laws, etc. Outside this entire web of relations – the whole *technostructure* – a car comes to grief about as quickly as a fish out of water.

Individual devices increasingly become points of connection to a network and while the network itself can be something physically substantial, as a rule it does not just provide a physical interconnection but an abstract set of links that divides society according to specific technical functions. It is thus wrong to talk about technology and its social consequences as if they were largely separate, for two main reasons: as we have argued in the last section, it is possible to think of the *very way that society organizes itself functionally* as a form of technification, and in addition, once in place, the technostructures of society are often used for new purposes by every new generation of users or as networks or terminals of new and different kinds. (This is, of course, not to deny that particular technologies – individual machines for instance – are often inseparable from particular social structures: the point has been made by Otto Ulrich,[16] and we would second it, above all to counter the illusion that a change to the means of production is possible without changes to the relations of production – the argument that exporting high-tech products and equipment is likely to have little effect on existing social relations within underdeveloped countries is clearly specious.)

The technostructures we have in mind are, of course, many and varied as well as being given to multiple uses. Their effects are bedazzling and virtually impossible to delimit – they have a way of spreading throughout the body of society almost of their own accord. Yet since contemporary society is part of technological civilization, it will no longer do to analyse it simply in terms of classes or social strata or organizations, or in terms of concepts of the state or the market or the family. The life of the greater body of society today is influenced decisively by these technostructures, while the life of the individual depends on his ability to function as a point of connection within the technostructures or to purchase the requisite connections within the technostructures.

In explaining the rationale behind the concept of a technostructure, we would emphasize again that the social meaning of technology is impossible to grasp simply through the concept of technological objects conceived as means to particular ends. Of course, the use of technologically more or less advanced tools solely as instrumental means is certainly possible; our point is that in an advanced technological civilization, such straightforward instrumentalism is no longer the norm. The conventional view of technology as a means seems to be drawn from simple examples: I eat with a knife and fork – are knife and fork not unambiguous means to an unambiguous end (getting my dinner into my mouth) and can technology not be understood on the same model? Yet even when we look a little closer at obvious, apparently trivial cases like this one, we find that from the point of view of social science, knives and forks are not just means for effecting instrumentally rational action. In fact, using particular eating utensils serves a purpose of its own, which is to establish a cultural distance between uncivilized gorging and civilized dining or, in other words, to *stylize* action in the sense required by the process of civilization.[17] The very example that was initially meant to demonstrate the simple sense in which technological objects can function as mere means in fact turns out to show the opposite, viz. that technological objects, even of the apparently simplest kind, themselves change the very form of social behaviour.

To argue that social relations or social behaviour exist in their own right and that human beings equip themselves with technological means merely to give material effect to such relations and behaviour is to gravely understate the social meaning of technology. The existence of technology is what makes some forms of human behaviour possible in the first place. This is not to deny that in most cases we can point to technological and pretechnological forms of the *same* activity – cases where technological change does not result in fundamental changes to existing anthropological or social forms. Yet it would be impossible to understand anything at all of the reality of today's highly technified lifeworlds if we insisted on thinking of a telephone call merely as a technically mediated form of conversation, or, say, car travel as a fast-paced form of walking, or photography as a more precise form of painting. Of course, we can explain the transformations of form involved in many such cases by interpreting technology as what Linde calls a social *institution*[18] – that is, as a social force which works to standardize particular forms of behaviour and set them on a permanent footing. Or, we can explain such transformations of form in terms of a transition from ethically standardized social action to instrumentally rational behaviour, as Habermas seeks to do.[19] In general though, we have to be very clear that the new forms of behaviour are *essentially technological* – that the behaviour in question is behaviour that has been shaped fundamentally by one or another technostructure.

Technological socialization, technological emancipation

Armed with the concept of technostructures, we can now risk making some sweeping pronouncements about what society is, and what it could potentially be, in the context of a technological civilization. Before we start though, it is worth recalling that far from everything human beings do – far from everything that happens to us or is relevant to our lives – is somehow quintessentially social. An argument of the form 'everything is social' might sound radical, but it is largely vacuous. The notion of society acquires definite contours precisely when some things are excluded from the domain of social relevance. Such was already the case for the Greeks, who made a radical distinction between the various degrees of citizenship and the *idiotes* (the person lacking a professional skill, the private citizen, or, more simply, the individual).

Clearly, where the line of demarcation lies between social and non-social domains has varied considerably throughout history. The bourgeois society of the eighteenth and nineteenth centuries famously gave rise to the idea of privacy (e.g. the private domain of family life) at the same time as creating a distinctly bourgeois public sphere. The implication was not that the family stood entirely beyond society, but that the family played its social role *en bloc*. In general, it was represented in the public domain by a male family head, otherwise what happened within the family was socially irrelevant.

Thus, what we have to come to terms with in the contemporary world is that technostructures give rise to new polarities of social assimilation and exclusion, creating new forces that push people and groups in and out of the social domain. In short, technostructures work equally to integrate, isolate *and* liberate people from the main body of society.

1. One of the fundamental questions that can be asked of any society is what shapes the mass of human beings who make it up into a single more or less cohesive group: what holds it together? Durkheim – to give one of the more notable sociological answers to the question – calls the relevant organizing principle *solidarity*, which he differentiates into two types. In the case of *mechanical solidarity*, human beings are bound together within the broad social whole by common beliefs or values. In the case of *organic solidarity*, the links are provided by multiple modes of work which supplement each other to produce a society which constitutes a different kind of whole, something resembling a worker whose many skills necessarily support one another. For Durkheim, the two forms of solidarity correspond roughly to the difference between traditional and modern societies.

My claim now is that we find ourselves at the beginning of a phase where technostructures are taking over the function of social integration. Numerous examples come to mind, our society's enormous networks of supply and waste disposal, or the so-called media of mass communication (the latter in particular seem almost specifically designed to provide a modern surrogate for the mechanical solidarity of old). Historically speaking, if the image of social integration was constituted in traditional societies through possession of a common culture, and in earlier modern societies by channelling the action of individual economic producers into common economic markets, today it is a technological form of integration that is taking shape alongside, and in competition with, those earlier integrating forces. What we are talking about is nothing less than a remarkable interdependency of practically everything that happens that is thoroughly mediated by technical networks.

Habermas has attempted to conceptualize these developments with the help of a dichotomy between systems and lifeworlds.[20] Yet although he introduces his distinction initially as a difference between *perspectives* – the external systemic perspective is contrasted with the perspective of the participants in lifeworlds[21] – he goes on to explain the difference as one between competing forms of integration: in the later portions of *The Theory of Communicative Action*, systems integration is contrasted with social integration.[22] Although the discussion centres on contemporary types of social structure, technology receives little more than cursory attention.[23]

However, if we make good Habermas' omission and restore technology to its proper place at the centre of the theoretical picture, it becomes clear that what Habermas calls systems integration amounts broadly to a form of *technological* integration and that it acts crucially to replace the social media of earlier phases of modernity, viz. money and power, with a new medium, information. Strictly speaking, what for Habermas goes by the name of social integration, viz. the linguistically mediated co-ordination of social action, is not really a form of integration that applies to the whole of society, since it can, at best, be said to apply throughout society in the form of technical networks (viz. communications networks), while integration in lifeworlds always remains on a local level. What Habermas provides us with is a confused picture, not of two competing forms of integration, but of a single technically mediated form of systems integration – something which could be described from the perspective of the participants, viz. the purchasers of technical services and expertise. On the other hand, we have the sphere of everyday communication within lifeworlds – something which our society's advancing systems integration seems to condemn to an increasingly marginal presence. Clearly, this marginalized aspect of social existence tends also to have subversive – perhaps one might say explosive – potential.[24] Yet we are still not dealing with a *conflict* between forms of social integration but at most with a dialectical *concurrence*, something like the reciprocal relationship between bureaucracy and corruption.

2. A second related question we can ask in this context relates to the form of individuals' social existence, or lack of social existence: how does it come about that somebody belongs to society or fails to belong? What possibilities or qualities enable the individual to participate in social life? In the societies of the past, property and labour were the prime criteria which decided how the individual was integrated into the broad social totality. Today, on the other hand, our social being tends to be that of a sequence of numerical codes or a point of technical connection; as individuals, we increasingly *come into existence socially* through one or another set of keys which provide access to various social networks. Plainly, that does not mean that someone who does not have the right sort of connections or credit cards ceases to exist entirely. No one evaporates into thin air; however, many cease to exist in a meaningful social sense. Nor is that something extraordinary or historically unprecedented: in bourgeois society, not every individual human being counted as a *person* – as we have just seen, only the male head of the family had this status because he was the representative of the family in the social sphere. In societies like the contemporary United States, it seems broadly true to say that someone who does not have a telephone no longer exists socially. With the gradual abolition of money, something similar is happening in the economic domain. If, until recently, purchasing goods with hard cash brought with it an act of exchange, of however symbolic a kind, what is set in train via a personal code during a credit card transaction is nothing more than the transfer of data. Money in the physical sense degenerates into a vehicle of regional intra-economic traffic between social nonentities.

In assessing these kinds of processes of social technification, we have to take into account that socialization implies a constant desocializing process and that the aggregation of social forms implies a concurrent splitting off of social forms. The future prospects conjured up by technified modes of social integration might in some ways be depressing. Yet the liberation of social irrelevance is something to be welcomed in certain senses as well. If one takes Marx's vision of future society as a model, then it admittedly looks like grim resignation to see the dis-integration of human beings as the most promising potential vehicle of the emancipatory project of old. Yet maybe the true liberation of the workers of the world could lie in a further technification of production to the point where work becomes socially superfluous. Likewise, maybe the contemporary rediscovery of the body is the flipside of the fact that large amounts of social action are set in train by codes and so no longer require human beings' physical presence.[25] Last but not least, maybe the extraordinary intensification of communication between human beings in the contemporary world is a reflection of the fact that communication is becoming more and more irrelevant in practical social terms.[26]

Societies or generations?

In trying to understand what society is in the age of technological civilization using the concept of a technostructure, we have obviously not been referring to society *tout court*, but to a form of social existence that is at present only just emerging. This means that the social structures of the past remain in place to a certain degree. Contemporary society builds on top of them; occasionally, tensions between the older structures and the new technostructures make themselves felt.

In this chapter, the picture of a technological civilization that we have put together so far has surely been overdrawn slightly – although perhaps still not overdrawn enough. As Stanislav Lem has realized, if one's aim is to devise a relevant social theory that provides some sort of compass for present and future social life, then an element of science fiction is unavoidable. Without it, our sociology will be a belated sociology, as it often was in the past.

It is precisely the influential role that technology plays in shaping society's basic structure that calls for a high degree of imagination in the construction of theories and for a dynamic approach to one's very concepts. With that in mind, we could even put a question mark over the argument of the present chapter so far. It comes in the form of a suggestion, viz. that in discussing a technological civilization, we might perhaps have been better off talking about *generations* instead of society. The concept of generations would have a two-fold advantage here: first, in bringing out the fact that our conclusions, like all conclusions in the social sciences, have their historical limitations, and, second, in allowing social life to be described from the point of view of the generation that is currently coming of age.

With both those points in mind, we would have to reformulate our argument as follows: the present generation of humanity is producing – over and above what is required for its own material reproduction – a vast quantity of waste, toxins and dissipated energy, as well as producing vast quantities of data and the technostructures of a technological civilization. Furthermore, just as the way we choose to use the environment will shape what is possible for coming generations of organic life, so the production of data and technostructures will shape what is possible for coming generations in terms of social life. In short, the raising of new technostructures will, in future, structure the field of social possibility and determine who or what is emancipated from social constraint in the sense of being consigned to social irrelevance.

Technological civilization and culture

Thus far, I have been attempting to pin down phenomenologically what the scientification and technological reconstruction of social life means, and it is tempting to bring together the various phenomena that have entered into our analysis by saying that technology will, in some sense, be the *culture* of the future. Technology indeed is coming to monopolize some of the important functions that were fulfilled by culture in traditional societies: the function of orienting action, the function of defining the social order, the function of forming society into a coherent whole – to name three rather momentous parts of the picture. Moreover, it is apparent that the worldwide advance of technological civilization has contributed substantially to the dissolution or collapse of traditional cultures. Traditional ways of life, traditional ways of interacting with the material world, traditional ways of solving problems, traditional forms of communication and movement, as well as traditional forms of production are increasingly detached from their original settings in the culture of particular regions and are subject to technical reconstruction. (It is not surprising in this sense that tourists today often have the impression that basic social structures are nearly identical everywhere.)

In fact, technological civilization stops short of completely destroying traditional cultures; it would be wrong to think of technology as simply abolishing traditional culture. However it certainly robs it of important functions. Once the general frame of orientation for all human existence, within modernity culture has become simply another sector of life: variously a sector of political life, a sector of production (what Adorno and Horkheimer call the Culture Industry), or else part of private life (a form of leisure on a par with many others).

The idea that a technological civilization will supply the culture of the new worlds of contemporary and future modernity is clearly an illusion. While it is true that technological civilization has a number of characteristics that are capable of spreading throughout all existing culture-worlds, technology remains something distinct from culture. A technological civilization creates a series of new tensions and dichotomies which push culture aside – marginalize it, without simply destroying it. That could mean that culture will become tendentially ever more irrelevant; however, it remains vital to clearly identify the tensions and dichotomies associated with the process so as not to overlook the *opportunities* that are being opened up at the same time. We have already noted the fact that the human body tends to become socially irrelevant within a technological civilization, and we have argued that that development contains within it the seeds of a possible rediscovery of the human body. Let me list a number of constitutive moments of the contemporary social world, all of which point ambiguously in the direction of both questionable and promising future developments within the broad cultural realm. A technological civilization, in broad terms, is bringing about a decoupling of:

- sex from reproduction
- eating from the basic necessities of physical nourishment
- imagination from perception
- thought from mechanical mathematical calculation
- movement from the organized flow of traffic[27]

Our examples show just what sort of Janus-faced phenomenon technological civilization is. On the one hand, we find life organized in strict, instrumentally rational terms. However, we also find the field of cultural developments opening up in previously unimaginable ways as a result. We can think of the ambivalence of the overall process on analogy with the formative moments of bourgeois society. The development of the bourgeois public sphere, which put an end to the arbitrary individual exercise of political authority by submitting everything and everyone to public forms of legitimation and the rule of law, simultaneously created the concept of bourgeois privacy: the intimate domain of family life. Similarly, a technological civilization is driving a process of sweeping social emancipation in the double sense of *making existing social structures redundant* and *opening up new fields of movement, imagination and free activity*.[28] The process, moreover, is radically self-sustaining. For example, new possibilities of movement in turn free up personal relations and individual consciousness as personal relations become superfluous to the normal functioning of society, and as many intellectual tasks involving particular types of thought come to be performed by machines.

It is the Janus-faced nature of technology that we need to keep clearly in view in formulating any definition of a technological civilization. Jacques Ellul has framed his definition in terms of the form of knowledge that he sees assuming a dominant role in developments.[29] For Ellul, the basic cast of a technological civilization is utilitarian: the notion of *efficiency* provides the basic pattern of thought in a technologically oriented world. Lewis Mumford's definition, on the other hand, emphasizes the most basic ontological feature of a technologically reconstructed world: in Mumford's language, that world is a world of *megastructures*[30] – or what we have called superstructures. However, in our view, neither Ellul nor Mumford have taken sufficient account of the ambivalence of the technifying process. Neither do the numerous writers who have paid particular attention to counter-cultures, alternative medicine, meditation, esoteric therapies and rituals, and sought to see in them a decisive *counterpoise* to technological civilization. In most cases, the advocates of alternative cultural and subcultural forms fail to recognize that such phenomena are nothing more than the obverse face of a single coin. Bringing both sides of the process of a technological civilization together, my preference would be for a different sort of definition: technological civilization, we could say, is *the very splitting off of instrumentally rational action from a humanly fulfilling life.*

Above all, we need to pay heed to the ambivalence of technification in making value judgements about technological civilization. Whether the profound impact of science and technology on our world is positive or negative depends not only on how we read the particular phenomena that we find inscribed on one or other side of the coin, but also on how we read them together.

After centuries of championing science and technology as major drivers of human progress, it is difficult today to assess where we stand without a degree of scepticism and foreboding. From the point of view of culture particularly, a deep-seated social technification in many ways looks like a backwards step. In trying to spell out the meaning of science and technology for our world under the heading of technological civilization, I have been suggesting that technology on the whole is neither inherently progressive nor regressive – that technification in a certain sense simply creates change. Further rapid development of science and technology in any case seems to be something like our historical fate. However, one thing we can certainly say about that historical fate is that at the end of the prodigious development that is now underway, what it means to be a human being will be something quite different from what it meant throughout the entire previous history of the world.

Knowledge society

Early hopes

In retrospect, the expectations associated with the concept of knowledge society[31] come across as curiously utopian. In the years immediately prior to and following

1968, they took the form of political hopes for a convergence between capitalism and socialism and for scientific solutions to the full gamut of human social problems: hopes for a definitive solution to the problem of world hunger, hopes for the abolition of factory labour or, indeed, for the disappearance of *all* heavy manual labour in the upwards sweep of technological progress. To speak with Marx, the realm of freedom seemed finally at hand, the realm of necessity looked like something humanity might be soon able to master. Perhaps the epitome of all these hopes was the so-called Green Revolution, the almost palpable sense that the progress of agricultural science and technology would finally make hunger a thing of the past. At the same time, the widespread automation of industrial processes promised by the products of the burgeoning sciences of cybernetics and electrical engineering held out the prospect of replacing labour itself with higher-order regulative or creative activity.

The general sense of possibility found its clearest expression when and where it was bound up with political idealism, perhaps most notably among the theorists of the Prague Spring. The collective headed by Radovan Richter[32] put forward the idea of a technoscientific revolution capable of putting a wholesale end to humanity's political and social ills. The sense that the progress of science and technology was synonymous with human progress, that its basis was something fundamentally *humane*,[33] was apt to make the difference between capitalism and socialism appear relatively minor, for in the end what science and technology hinged on were intellectual resources, and societies on both sides of the Cold War divide could be seen as contributing equally to their growth. The idea that *intellectuals* would be the leading class of the future flourished both East and West of the Berlin Wall. Books like Gouldner's *The Future of Intellectuals and the Rise of the New Class* and Konrád/Szelényi's *Intellectuals on the Road to Class Power*[34] – both of which run representative versions of the convergence thesis – were widely read and much discussed. Daniel Bell's more cautious vision of the future looked forward to the end of the industrial era.[35] Yet what Bell's soberest sociology saw arising to take the place of industrialism – a society whose economic base was a flourishing services sector – was still shot through with hopes that we cannot help regarding as utopian today. For Bell, theoretical knowledge would be the axis on which society would henceforth turn. Once they had made the great leap beyond a work-oriented society, human beings would use their new-found freedom to participate in all manner of social projects; Bell's work contains an unmistakable trace of the, for us, fond notion that a *truly* political age is about to dawn.

Today, it is painful to look back at the hopes that suffused the period around 1968, and the actual course that history was to follow, well to one side of the projections of Bell and so many others, has necessarily led to major revisions of the 1960s' original suite of ideas – such was the intellectual effect of continued world hunger, the rapid loss of agricultural land, the continued existence of poverty and exploitation, the growing use of technological progress for military purposes, the political passivity of the scientific community and the collapse of active democratic participation. (My own book, *At the End of the Baconian Age*, I now realize, was a small part of this sweeping critical revision.) However, what is necessary today is something different from a renewal or a repetition of by now widely voiced criticisms. Now is the time for us to take stock.

The history and present of knowledge society

Since World War II, the level of a nation's scientific and technological development has been one of the key measures of economic success and military power. In the industrialized world, research and development became a sector of politics in their own right and were allocated a separate department with a sizeable budget. Partly in consequence, partly as a result of broader social change, the education sector underwent dramatic expansion, with tertiary education leading the way. Between the mid-1960s and the mid-1980s, public spending on education and training came to account for around 15 per cent of government expenditure in many parts of the developed world, while the percentage of school leavers enrolling in tertiary education more than doubled; in excess of 20 per cent of high school graduates were moving on to universities or tertiary technical colleges by the end of the period, and subsequent decades have seen further rises. The figures are even more impressive when the growth of the working population is taken into account. The huge increase in the number of students attending universities and technical colleges led to a change in the ratio of those engaged in work versus study: while workers outnumbered students 3:1 in 1960, 20 years later the ratio had dropped dramatically to just 2:1. The expansion of course content and increased demands on students, together with the challenges of creating a genuine system of mass education, led quite rapidly to a situation in which education came to take up a longer and longer stretch of the individual's life-history. In Germany, it became the norm for students to complete their first degree as late as their late twenties, while the average age of doctoral graduates reached 32. Schools in general, and universities and research institutes in particular, thus came to constitute an enormous section of the overall workforce, a veritable social subsystem with major economic and political interests of its own. The production and reproduction of the knowledge involved – what we could call in bland terms the output of information – by now makes up a hefty proportion of GDP. Nor is the significance of contemporary education to be measured in quantitative terms alone. Qualitatively, education has become one of the leading sectors of the economy because of the dependence of so much production on expertise of one kind or another; everything from the pharmaceutical industry to communications and computer technology is part of the knowledge economy, or depends on it for ideas and trained personnel. Knowledge, in short, has become central to economic life because it is the central factor determining standards of technological development and product innovation and makes a decisive difference to a nation's ability to compete in the international marketplace.

The contemporary position

The concept of a knowledge society can no longer be used to pick out an era of social development. Though the term does indeed cast light on contemporary society from a certain perspective, it no longer describes a trend within overall developments. Here is the problem: it was merely as the conceptual marker of an overall tendency that the notion of a knowledge society marked off a definite era: its value was in articulating the close of a different era – what many have called the industrial age. Today our situation is different however: knowledge society is plainly a social reality. But what that means is that the tendencies that it gave rise to have come to a halt. Moreover, with the expansive phase of modern knowledge society at an end, what we can expect to follow are a number of significant structural modifications – a necessary process of reorganization that will shift the very relationship between society and knowledge.

1 De Solla Price predicted, as early as 1963, that the growth of science would falter as it came up against external limits in the form of limited financial or intellectual resources.[36] European science today already finds itself in a phase of zero growth, with the conclusion of the Cold War arms race expected to bring about a further drop in activity within the research and development sector. The process will doubtless be delayed due to institutional inertia (in the case of defence research, by arguments for rearmament in the wake of current disarmament). However, the effect of disarmament within the scientific base is unlikely to be offset by increased research output in other fields. (The outlook in other parts of the world is undoubtedly different, not least because of intense interest in military research – and military activity – in the opening decade of the new millennium.)

2 Rescher's 1978 studies of scientific progress come to the conclusion[37] that the law of marginal utility applies in science too: the resources necessary for achieving positive results increase steadily over time. Rescher goes on to argue that science will come up against absolute limits, not so much in analysing complex relations, but in addressing problems of large-scale synthesis whose solution calls for one-off, capital-intensive use of resources (for instance, in astrophysics and elementary particle physics).

3 The growth of tertiary education has come to a halt – indeed, the sector as a whole has shrunk due to the fiscal difficulties facing governments in many

parts of the world. In the near term, the German university system faces annual cuts of up to 30 per cent, including staff cuts of up to 5 per cent.

4   Among the so-called end users of academic expertise, we see some signs of resistance to the dominance of theoretical forms of knowledge. Industry occasionally shows a preference for college-educated engineers over university graduates, while growing numbers of patients are turning to alternative therapies instead of academically certified members of the medical profession.

Critical social theory, hierarchies of knowledge

In response to these kinds of development, the theory of knowledge society is shifting increasingly in the direction of critical social theory:

1   In contrast to Bell's theory of post-industrial society[38] and Krebich's theory of scientific society,[39] critical social theory attempts to characterize contemporary society not just from the point of view of a *dominant* form of knowledge, but in terms of *multiple forms* of knowledge and their reciprocal relations and future potential.

2   In contrast to theories of information society, critical social theory pays serious attention to the *subjective* aspect of knowledge. Knowledge is recognized as the knowledge *of a particular human being*. Information is seen as something that has multiple potential meanings.

Perhaps the crucial point here is that the concept of knowledge that is central to critical social theory takes in all possible forms of knowledge, from implicit knowledge (tacit knowledge, know-how, etc.) to the explicit knowledge that we find methodically sharpened and theoretically developed within science. As a conceptual tool of critical analysis applied to contemporary society, such a notion of knowledge makes it possible to study the way that interrelations between the various forms of knowledge are shaped by the relationships between knowledge carriers and between the social groups that they form (what Znaniecki has dubbed "*social circles*"[40]). Interpreted as a form of cultural capital, knowledge comes to be seen as a factor in power relations, as a potent force driving social development, and as a fundamental determinant of life chances.

In relation to the crucial issue of knowledge and power, it is important to recognize firstly that hierarchies form (and re-form) between the various types of knowledge, second that barriers and exclusionary strategies operate to restrict access to knowledge, and third that such barriers and strategies exist, in part, to underwrite the status and influence of the carriers of particular kinds of knowledge. The hierarchic division that is probably most characteristic of contemporary society is the gulf separating scientific knowledge and knowledge derived from everyday human lifeworlds; of all the various forms of knowledge, scientific or academic knowledge occupies a position of social pre-eminence.

Though its historical derivation in philosophical *truth-seeking* suggests that it is a completely general form of knowledge, scientific knowledge is, in practice, usually limited to circles of experts: education and the limited accessibility of the tools of the scientific trade – everything from raw data to concepts to instrumentation – represent practically insuperable barriers to general understanding. In addition, there is the fact that the exercise of scientific expertise depends on institutional authorization, that is, certification in the form of a licence, diploma or other degree. Forms of work based on academic knowledge are generally denoted as *professions* and in today's world what we see is that the social respect accorded to professions has increased to the point where many non-professional forms of work spare no effort to acquire professional status. Some fields attempt to make work in the relevant field conditional on academic training; others slip into a state of passive dependence on academic knowledge and its positively privileged carriers.

The caring professions

The classic example of these developments is in the field of medical care, where carers of various types (nurses, midwives, etc.) have gone down the path of professionalization or have either been forced to cede their original independence to technically trained medical professionals. The result is two-fold: on the one hand, in the course of professionalization, carers tend to lose most of what set them apart from medical science – usually a knowledge grounded firmly in practice (what Michael Polanyi has called *tacit knowledge*). On the other hand, the gulf between laypeople and scientifically trained experts continues to widen. In the medical sector, there is the additional problem that attempts to demarcate or close off scientific medicine from other forms of medical practice has led to the repudiation of all non-European medicine – a development that in sociological terms becomes entrenched through the influence of doctors within the health and insurance systems. The rapidly increasing costs of health care, though sometimes exaggerated, are in no small part a consequence of the elimination or suppression of *intermediary* forms of medical care; even a sore throat or a light fever these days calls for a trip to a doctor, whose expertise is the product of years of costly training and is underwritten by a football field's worth of medical equipment that he is probably hard at work paying off.

Technification in the field of medicine demonstrates how the concept of knowledge society can be used to cast a general critical light on professional hierarchies.

Two vital questions arise: first, whether the dominant position of scientific knowledge within contemporary society is leading to a decay of other forms of knowledge, and, second, whether there are types of problems, and types of human need, that scientific or technical knowledge is particularly ill suited to address.

The dominant position of technoscientific knowledge in contemporary society implies a belief that this particular form of knowledge has the last say in establishing all matters of fact, and in solving problems of all kinds. Scientists are regularly hired by public interest and citizens' advocacy groups to generate a degree of general awareness of problems that might otherwise be neglected. However justified these groups' concerns may be, the social landscape in which they represent them is marked by the gradual and equally justified recognition among the public at large that scientific reports do not definitively settle complex problems of social policy or social action, and that such reports often reflect the points of view of the groups that commission them. However, the authority to frame problems – the ability to speak about problems in a relevant way – remains everywhere the sole prerogative of professional groups.

Experts and laymen

In basic terms, the fact that human beings are social beings, living in societies, implies a division of labour and the exchange of goods. However, our contemporary knowledge society, as it has come into existence in recent decades, is characterized, on the one hand, by the intensive professionalization of most labour and, on the other hand, by the ever more intensive exchange of services. What that means is not just that modern human beings are dependent on others for access to the material goods that they are unable to produce themselves, but also that they are forced to entrust major aspects of their lives to experts. Education, care for the body and soul, conflict resolution, communication problems, relationship problems – all become the prerogative of technical or quasi-technical specialists to the point where the individual denizen of today's knowledge societies falls into a new sort of dependence on experts or what might be called a new state of cognitive and existential tutelage. We could note here that one of the truly representative works of the late Enlightenment – Kant's famous essay of 1783 'What is Enlightenment?'[41] – no longer attributes human beings' self-imposed immaturity (*Unmündigkeit*) to dependence on the church or the state, but to dependence on *experts* (teachers, doctors, spiritual guides, etc.). Today, the situation has led to the formation of a variety of counter-movements that aim to bolster the individual's powers of self-determination and general competence vis-à-vis experts and expert bodies. Most achieve widespread currency in the form of catchwords: patients' rights, death with dignity, home schooling, parental prerogatives, consumer rights, etc.

Production and social reproduction

The production and reproduction of knowledge is fast becoming the most important and perhaps the largest sector of economic life, a situation which calls for us to consider the peculiarities of treating knowledge as a commodity. Unlike goods of other kinds, knowledge remains in the possession of the individual or corporation which sells it, even after it has been sold; further, knowledge is not something to which the conventional concept of consumption even remotely applies: making use of knowledge does not diminish the stock of knowledge. That, in turn, has major consequences in that the diffusion and use of knowledge are subject to artificial constraints – confidentiality provisions, copyright law, patents, intellectual property rights, etc. Such regulations run counter in a certain sense to the very essence of knowledge – or at least to the essence of *scientific* knowledge – which aspires to be universally valid and widely accessible within the public domain. Contemporary knowledge society is thus marked by an inner tension between knowledge considered as a public good or as common cultural capital, and the knowledge economy, which relies on the privatization of knowledge. The tension is detectable in many corners of society: it shapes the development of the internet (or rather the possible uses of the internet); it creates tangible political conflict where researchers or corporations seek to patent genetically modified species; and it affects all areas of social life where data protection becomes a (real or perceived) matter of urgent necessity.

Human societies exist as cultural communities only as long as the knowledge they produce is not the exclusive preserve of individuals but of wider social groups: the knowledge of cultural communities is given a long-term *intersubjective* foundation in a process involving symbolic representation and the formation of rules (this, we might say, is precisely how knowledge *becomes* cultural capital). The problems associated with the process show up in the contemporary world in various forms:

1   Individuals are no longer obliged to acquire knowledge themselves; indeed, as a result of the general accessibility of existing bodies of knowledge, they are liable to experience a dizzying expansion of their horizons, as the possibilities of action and the possibilities of access to different kinds of objects and mental universes open up. Yet dealing with bodies of knowledge that circulate more or less freely in the form of social capital requires learning: in brief, it means endowing symbolic representations with meaning, becoming competent in applying relevant rules and manipulating relevant techniques. That, in turn, leads to a fundamental change in the nature of learning. One's induction into a knowledge society centres on the acquisition of the knowledge of rule-based systems and patterns of cultural significance rather than knowledge of facts.

2 Because the individual's prospects within a knowledge society depend largely on his ability to participate in the formation and use of cultural capital, the increasing volume of such capital means that the proportion of his life devoted to education is constantly rising. As the growth of cultural capital continues apace, an end to the trend is not yet in sight.

3 Stable bodies of knowledge also require ongoing *reproduction*. Even where social capital takes on a symbolically fixed form to the point where it can be stored *en masse* in the form of data, the reproduction of such fixed forms poses an enormous problem (aside from the ongoing problem that individuals face in imbuing symbolic representations with some sort of ongoing living meaning). Contemporary knowledge societies are dotted with libraries, museums, data storage systems and what are fashionably called knowledge silos, all of which call for continuous conservation, reorganization and reproduction.

Since knowledge that takes the form of cultural capital is, in principle, something general, the economic, military and social power that knowledge bestows is no longer based on knowledge as such, but on the creation of *new* knowledge. The power dynamics of the knowledge economy provide a major spur to innovation and to the production of new knowledge, as we will see next.

Knowledge and power

During the Cold War, there was a widespread view that military superiority was essentially a matter of scientific and technical superiority, and, in particular, that such superiority was not based on knowledge per se, but on cutting-edge knowledge. The universal applicability of technoscientific knowledge meant, however, that the competition taking place at the forefront of knowledge simply raised the capacity of both sides to wage war, in the process raising the terror of war to almost immeasurable heights.

Over time, the military significance of knowledge has led to the strict classification of scientific research and unprecedented levels of secrecy. However, defence research is not the only field caught up in such developments. A country's technoscientific capacity is also crucial to its economic success and international competitiveness. Companies and nations unable to hold their own in the technoscientific struggle for survival are condemned sooner or later to dependence on other companies and nations; either they are obliged to become supply firms (or supply economies) or else their own production becomes dependent on costly licences dispensed in technically more advanced parts of the world.

Science and technology have become the engine room of national and corporate potential and they are eclipsing other traditional factors in the struggle for power – including money, as the relationship between the industrialized Western nations and countries like Saudi Arabia plainly shows. In the absence of scientific and technical know-how, vast capital reserves do not translate into political power. In political terms, the economic divide between the geographic North and South may indeed derive from the superior economic productive capacity of the North. But the latter, in turn, is derived from the North's superior technoscientific capacity. The relocation of industry to the developing world will do little in itself to cure the South of its dependence on the scientifically advanced nations of the wealthier parts of the globe.

Knowledge and opportunity

It is a well-known fact that the social status of the individual within a contemporary knowledge society is largely dependent on educational attainment. If, according to the sociologist Talcott Parsons, one of the distinctive features of modern societies is that modern social status is *achieved not ascribed*, what that means in practice is that social position and social power depend largely on educational status. Clearly, the traditional criteria of status, viz. wealth and birth, remain important. But they are by no means the dominant factors any more: the status claims of wealth and birth can commonly only be made good via education and training. This situation has led to a general push to professionalize and to pursue ever higher qualifications; it is the main contributor to the problem of overqualification, and to a relative oversupply of academic skills in the labour market. Whether we can talk of a veritable academic proletariat seems debatable. Certainly, there is no ignoring the other broad consequences of current trends to hyper-education though – for example, the way established professions raise educational entry barriers or attempt to expand their activities in order to create work for ever larger armies of qualified practitioners.

Yet it would be wrong to think of knowledge as the source of opportunity only when it takes the form of professional skills and professional status. If we define knowledge as cultural capital, it becomes clear that the doors open to the individual in a contemporary knowledge society depend on his ability to participate in the creation and circulation of such capital. The wealth of the developed world is by no means mere material wealth; it is also wealth in a cultural sense. As the material barriers to cultural resources continue to fall, for instance as information becomes more widely and cheaply accessible, the individual's opportunities to participate in social wealth become an independent source of opportunity. The effect is amplified by the fact that the individual's working life makes up a falling proportion of his total lifespan. Holiday and evening courses as well as the phenomenon of the mature-age student take on an increased significance both quantitatively and qualitatively. The average student's overextended years of study are often no longer truly a preparation for later life. Study comes to be *experienced* as a self-contained phase of individual existence.

Education thus becomes an independent mode of individual fulfilment. More importantly, it comes to be an independent mode of social *integration*. Because work functions less and less as a force for social cohesion within knowledge societies – to the point where some theorists foresee the not too distant end of labour-based society – it becomes politically counterproductive to make extra cuts to the education sector in times of high unemployment. In brute economic terms, the cost of places in universities and colleges is a tiny fraction of the cost of creating new jobs, while scrapping student places adds to the ranks of the unemployed.

Socialization through knowledge

The way knowledge has come to saturate society leaves society anything but unchanged. Just as applying natural science to the natural world depends on precon-ditioning nature (which in the final analysis means placing it technically at science's disposal in a purified, isolated form under controlled conditions), so progress within the social sciences brings little real benefit unless individual human beings act according to scientifically well-defined norms, rather than in accordance with their relatively opaque natural inclinations. As we have seen, Max Weber conceived of this situation as a relationship between the rationalization of social relations and the iron cage of modern life. In knowledge societies, what the relationship implies is that individual behaviour is continually forced into narrower, more tightly controlled channels; life presents people with more and more starkly defined alternatives. As a rule, however, the tightening of social controls is *not* brought about through educa-tion, as it was in the eighteenth and nineteenth centuries, but by purely technical means. In this sense, the first generation of voice recognition devices – which train their owners in the use of highly articulated speech made up of clearly defined terms and simple alternatives – might be a precursor of things to come.

Saturating the social fabric with scientific and quasi-scientific forms of knowledge also leads to a profusion of data, and to the exercise of social control through the processing of data. As a result, society certainly becomes more transparent in a sense. However, the greater the amount of social regulation that takes place via data management and data processing, the more people and things only count as part of society to the extent that they can be defined in terms of data. The increasing flexibility of working hours and the overall trend towards individualization, e.g. of product lines and product choices, have a counterweight in ever stricter and more numerous registration procedures and in the ever tighter controls exercised by means of the data that individuals produce. Detailed monitoring of every aspect of work, the multiplication of quality controls, the increasing importance attached to test results measuring psychological disposition, performance potential, health risks, etc. – up to and including genetic mapping – these indeed are the threads of the invisible web in which people finds themselves entangled in a knowledge society. One of the upshots of that situation is the barely manageable problem of data protection. The other side of the same problem is the way existing data about the individual can substantially diminish individual freedom: data are, in some sense, taking over the traditional role of fate. Extreme versions of that worrying predicament can be imagined in a world in which genetics and genetic technology continue their rapid advance. Such a world would call for a specific *right to non-knowledge* – a veritable right to ignorance of one's own genetic make-up.

Because of the way work is gradually ceasing to define life at the end of the industrial age, contemporary knowledge societies tend to produce more and more socially marginalized groups, in spite of the growing wealth of society as a whole. Having made the integration of the individual as part of local paternalistic social structures a thing of the past, the newly forming industrial societies of the nineteenth and twentieth centuries looked to accomplish a new kind of social integration via the market and the world of work. Today, however, neither the market nor work suffice to provide patterns of social integration. One might call it the central problem of a knowledge society – the problem of finding a new principle of integration that is capable of bringing the individual into meaningful connection with the social whole. Unless political efforts are made to engage a broad public in informed discussion and decision making, then there is a manifest danger that data management will take on the function of social integration by default. Examples from the contemporary world need only be varied slightly if one wants to think through the possibilities here: in Sweden, social existence is no longer possible unless one has a state-issued personal identity number. Likewise, in the United States, being a member of society is impossible in a meaningful sense without a social security number.

Knowledge society: Future prospects

Knowledge society is no longer a notion to which hopes of social or human progress are widely attached; if anything, it would appear that the expansive phase of the history of knowledge society has come to a close, and that the basic structures of knowledge society have become the cause of social problems of their own. No doubt, the internal dynamics of knowledge society are still operative: mountains of data continue to accumulate, the saturation of society with knowledge and its associated products continues unabated, productivity continues to grow as manual and non-manual labour is performed by computers or other machines.

At the same time, various disturbances make their presence felt:

- there are clearly limits, both institutional and intellectual, to the growth of science (military disarmament in the aftermath of the Cold War underlines the fact that in several parts of the system significant overcapacities exist)
- higher education is no longer a guarantee of social status
- universities are suffering from a decline in standards
- the belief that all human problems can be solved by scientific means appears more and more questionable

Nonetheless, it would be remiss not to mention some of the more hopeful prospects that could potentially open up if efforts were made to address some of these problems:

- An expansion of the cognitive, philosophical and experiential basis of the medical profession is conceivable. So too is a general rehabilitation of traditional vocations and their associated bodies of implicit knowledge.
- Similarly, a revaluation of the relationship between education and personal development – between the process of learning and the cultivation of the human personality – looks possible, as does a recalibration of the relationship between education and work.
- Finally, there is the discernible prospect – as dim as it sometimes seems – of a reinvigorated Enlightenment. In practical terms, that would require new public discourses capable of sensibly regulating the constant growth of knowledge, together with the emergence within the public sphere of a more self-aware and better informed core of citizens that is capable of (a) giving specific shape to social existence in a non-labour-based society, (b) asserting its independence from expert authority, and (c) resisting the impositions of data management and data control in the name of a redefined ideal of human freedom.

Trust in modernity

Talking about trust

There is something suspicious about talking about trust: the very act of doing so seems to imply that it – or *something* – has already broken down, that what we took as given has disappeared, or is on the verge of disappearing. Is it true that our trust in a person (or a situation or an institution or a thing) lasts only as long as that trust goes unspoken? Sometimes it seems so, for *professing* one's trust is certainly paradoxical. In short, taking trust as given is a basic presupposition of our dealings with the world, while talking about it normally only becomes necessary when it no longer exists – and in such situations, talking further undermines the presupposition of talk itself.

Daily life presents us with examples of our paradox that will be recognizable to everyone: 'You've just got to trust me!' is the common formula with which we react to a partner's loss of trust, something that I might say to implore and, at the same time, challenge the other, and that can easily put an end to what is being implored, for in *demanding* trust I can easily awaken the suspicion that trust is particularly called for, or that I would prefer to be given *carte blanche* and avoid all further discussion.

If raising trust as a topic for general reflection about our society is thus, by default, to concede that something is already amiss in today's social world, it is important to point out that it is only one type of trust that puts us in this paradoxical position. Psychologists, following the lead of Erik H. Erikson,[42] have called it *Urvertrauen* – a kind of trust that is primary or primordial in the sense that it precedes all speech. As a psychological concept, primary trust denotes a capacity that develops in earliest childhood and stems from the sensation of being in good hands or surrounded with care. Nonetheless, primary trust is not the term that I will use in what follows; it makes trust into too much of a matter of individual psychological development. If it is important to raise trust explicitly as a *social* issue, then treating it as a problem of individual psychology would seem to entail taking a secondary problematic for the primary one, for the disquieting loss of trust that is manifest in the contemporary world is, first and foremost, a sociological phenomenon. The type of trust that I want to think about is something that I will refer to instead as *ontological trust* – coined on analogy with Anthony Giddens' notion of *ontological security*.[43]

Whether we call it primary or ontological, we need to distinguish this kind of trust from the kind that is essentially a personal accomplishment, or, in classical terms, a *virtue*. Like primary trust, the virtue of trust is something I can never demand of the other, although I certainly demand it of myself. Unlike primary trust, the virtue of trust is not a psychological resource that has its roots in childhood, but a kind of ability that can be learnt and practised by any mature adult human being. Trust as a virtue presupposes a knowledge of life's dangers and a first-hand experience of the shortcomings and unreliability of other human beings. It is quite different from a conscious feeling of security. Perhaps its most characteristic mark is a readiness to expose oneself to the dangers of the world and the perils of possible hurt at the hands of others. The virtue of trust is an ability that distinguishes the sovereign human being.[44]

Obviously, there can be no question here of blindly courting danger or putting oneself in the hands of anyone at all, let alone at the mercy of those one knows to be unreliable. We could say that this kind of trust is a certain steadfastness, one that presumes on the basis of experience, and in the absence of indications to the contrary, that relationships will live up to what they promise and that the people one normally associates with are trustworthy. Trust as a virtue has a certain similarity to the sort of courage that involves an unshaken consciousness of what one does and does not have good reason to fear.[45] (We recall that Socrates already distinguished courage from audacity.) Trust as a virtue is also related, in a certain sense, to the classical virtue of magnanimity insofar as it always involves a presupposition of good will on the part of those that one interacts with. As little as it can be demanded, this kind of trust is *itself* a kind of demand: trust in this sense *calls for* trust.

The loss of trust in modern society

It is striking that discussions of trust arise at crucial points in so many theories of modern society, perhaps because theorists habitually associate trust with traditional forms of social life and read modernity in terms of its very opposite – the growth of norms and controls to replace the need for trust. In this vein, Weber thinks of traditional societies as communities based on high levels of unspoken reciprocal *consent* among community members.[46] *Modern* trust, in Weberian terms, is something quite different – a form of socialization grounded in notions of rationality, i.e. in norms, controls and explicit rules governing behaviour. Weber can still be read as arguing that societies (in the strict sociological sense) have their antecedents in communities and in communal forms of consent. (And, indeed, this very question is not without relevance to contemporary debates about communitarianism.) Yet how is it that trust also becomes an issue in social theories like Luhmann's, which present themselves emphatically as theories of *modern* society? The answer will be obvious to the reader of Luhmann's important 1968 work *Trust*: it has to do with what is new about Luhmann's theory, which is that he thinks of trust *functionally*, like so many other facets of human life, with the result that he recapitulates one of the prime tendencies of modern society – the thoroughgoing rationalization of all social relations – at the level of theory: here, trust too has become a means to an end. Luhmann's theme in *Trust* is 'the question of the maintenance of a broad range of systems of human action'.[47] With the question of systems of action in mind, Luhmann would have us ask – what does trust achieve? The answer is that it *reduces complexity*.

Now we can readily agree with Luhmann that from the point of view of 'human systems of action' (i.e. systems made up of human beings engaged in functional activity), reducing the complexity of modern societies is something highly necessary and desirable. As Luhmann sets out the problem, the main question seems to be whether trust is really the most intelligent means of effecting the required reduction in complexity. Faced with the problem of growing complexity, we could after all adopt the (patently superficial) policy of simply shutting our eyes and pressing ahead regardless. Luhmann, however, convincingly shows that the alternative to trust, viz. mistrust, though also a means of reducing complexity, is a far more costly and inflexible policy, and in the end one that tends to self-fulfilling prophecies: it creates the very dangers it lives in fear of.

An even more fundamental objection to Luhmann, however, is that he assumes that the world is more complex than individual human beings *qua* individual systems of action. Nor is that by any means a trivial assumption; it seems to rest on the old-fashioned view that large systems are necessarily more complex than small systems. Yet modern natural science has plainly taught us otherwise: for example, it has shown that the human brain is almost literally a system of astronomical complexity – that it may indeed be equal to the task of coping with any system of a smaller degree of complexity than the entire universe. Does it make a difference here that Luhmann is talking specifically about modern societies, and not about traditional agricultural societies or the relatively small, self-contained bourgeois societies of the eighteenth century? Modern societies are indeed considerably more complex than that (there is no denying that for the individual they are simply *too complex*), though we need to add here that this complexity is something directly produced – and here my argument diverges from Luhmann's – *precisely in order to compensate for a loss of trust*.

Why is it, we might well ask, that contemporary social life is subjected endlessly to quantitative analysis? What purpose is served by the contemporary profusion of general norms and detailed regulations, its swarming multiplicity of monitoring procedures, performance evaluations, surveys, checks and balances? The general answer is that we hope to attain through managerial control what we no longer trust *in trust itself* to bring about. To take just one example: when one no longer trusts the intuitive diagnosis of a GP, one has no option but to submit oneself to a string of tests and analytic procedures. It seems inadequate, in other words, to simply take the complexity of the contemporary world as a given, as Luhmann does. What we are dealing with here is more like a vicious circle that is actually *driving* the development of modernity. The general loss of trust is leading to an increase in complexity, the reduction of which supposedly depends on – more trust.

What becomes clear is that, as a characteristic mark of modern society, trust as Luhmann conceives it is something like second-order trust, the trust as it were of a world in which trust is a thing of the past, or, more specifically, a world in which primary or ontological trust is a thing of the past. To see why that might be the case, we can turn to a more recent theory of modern society. As in Luhmann, in the work of Anthony Giddens we get a strong sense that the modern form of trust, although differing from primary trust in being markedly more risk conscious, can in no way be equated with the classical virtue of trust. Giddens' modern form of trust is itself only a kind of convention. We could call it a social institution, but certainly not a personal accomplishment.

On Giddens' theory, modern societies are distinguished by a quality that he labels *disembeddedness*. What that means essentially is that the modern individual no longer comes into contact with the basic conditions of his life and is thus no longer in a position to put his trust in them. In general people neither have direct access to the sources of the food them eat, nor do they possess the kind of knowledge that they depend on to sustain the web of their material existence, nor can they influence the operation of the various large-scale technological systems that they make use of. Similar dynamics shape modern individuals' relation to both politics and the economy. Trust, according to Giddens, is thus needed above all to compensate for individuals' ignorance of the workings of the world around them and the practical impossibility of coming into direct contact with the material basis of their own existence. And it is central above all to their way of relating to Giddens' so-called *expert systems*. What he has in mind with the latter concept are broad social institutions like medicine and law and the general sphere of technics (IT, communications, computing, etc.); we typically confront these at what Giddens calls *access points* – what are, in effect, the social sites where individual experts present themselves as representatives of their respective systems.

Giddens thus sees the relationship between experts and laymen as the heart of modern trust. Here, too, trust turns out to be the very opposite of a personal accomplishment; trust, in Giddens' sense, is simply a sort of social convention, however much we might each individually experience it at times as an unreasonable imposition. If it is an imposition, that is above all because evincing such trust means delegating one or another aspect of one's life to experts and thus submitting more or less wholly to expert *authority*. We should again recall here that at the close of the Enlightenment – *not* in its earlier phase – Kant defined Enlightenment as human beings' emergence from a self-imposed state of tutelage, and that our self-imposed tutelage is something that he identifies precisely as a *dependence on experts*.

There can be no doubt that the type of trust that we are talking about – second-order trust, trust as a social convention – is a defining feature of modern societies, as well as a means of reducing complexity. The complexity created by the subsystems of modern society is indeed not something individuals can cope with on their own. It is barely possible to represent oneself in court, or even buy a house without the help of a solicitor of some sort. As a patient seeking medical treatment and trying to exercise a degree of informed consent, it is impossible to form an independent opinion of the data on which a doctor will base his recommendation to proceed with a particular type of therapy. Nor can one have the remotest idea of what kind of environmental hazards, ambient radiation or heavy metals that one is exposing oneself to, nor which cocktail of vitamins is actually going to be beneficial to one's health. Self-determination, Enlightenment-style, is indeed a fine ideal. But life in modern society seems to demand nothing but trust, trust and more trust.

Ontological trust

The type of trust that is usual in modern society gives us whatever fragile confidence it does because it is something *conventional*. At the cost of reducing the individual to a state of unenlightened dependence, it provides us with a precarious defence against some very deep uncertainties. Yet with the emergence of this new form of trust, a far more fundamental form has fallen away – primary trust, or what I would prefer to call ontological trust. But what was ontological trust to begin with? And what are the signs that we have lost it?

Let me turn first to trust in nature, or more specifically, human beings' bodily nature, the nature *that we ourselves are*. Ontological trust in this sense is something we are simply unable to live without. Nonetheless, it has its acute fracture points in today's world; all in all, it seems to be on the wane. Trust in this sense is trust in the fact that practically everything to do with our physical being happens of its own accord, trust that our bodies will essentially do what we expect of them, that they will digest what we feed ourselves, that wounds will heal, that our hearts will keep beating regularly, that we will fall asleep without further ado at night, that our sexual appetite (and ability) will not desert us, and so on. The list of what we take as given becomes very long indeed – frighteningly so in that every item contains within it the shadowy possibility that nature could withdraw its spontaneous powers of action. One thinks of the traditional German verse: 'Morgen früh, wenn Gott will, wirst du wieder geweckt' – 'In the morning you'll awaken if God pleases'. No doubt, it was intended to be unsettling, to make children frighteningly aware that they go to sleep *blindly trusting* that they will wake up the next morning.

The doubt that what makes up the very fabric of our being might cease to happen of its own accord can take root at any point. Yet we don't have to stand by helplessly. After all, we have the medical means of coping ready to hand: sleeping pills, stimulants, sedatives, anti-impotence pills, digestive aids – all effective weapons against our underlying anxiety. Who could doubt the general effectiveness of the many instruments that medical technology places at our disposal? What is questionable though is whether their use undermines our trust that our bodies will spontaneously do what medicine is supposed to induce them to do by artificial means.

As a second locus of ontological trust, let us take trust in external nature, the age-old faith that the sun will rise in the morning, that spring will follow winter, that the earth will continue to nourish us, that the soil will remain fertile, crops edible and water drinkable. That faith has been severely shaken (or, to put it in positive terms, a growing *environmental awareness* has been a major feature of the social history of the past half-century). Tellingly, the worldwide environment movement was set in motion by the fear sparked by Rachel Carson's *The Silent Spring*.[48] The book's title suggested a picture of spring without birdsong. What it set in train was a certain sort of sociohistorical unmasking. Suddenly, it became clear that up to that point in history human beings had simply *assumed* that the world of nature would always continue to exist independently of our collective action as a species. In his poem 'Epilogue to 1949', Gottfried Benn had still been able to write:

und Schmetterlinge, März bis Sommerende,
das wird noch lange sein
(Butterflies from March till summer's end/so the world will go for a long time to come.)

Today, nobody would dream of penning such naiveties. Within a decade or two, what the trust that sustained us in the past actually amounted to was plain for all to see: something like the trust of children who behave as badly as they please towards their parents because they are unable to imagine that anything they do could possibly endanger their parents' love. Nature in this sense is almost the epitome of what sustains us: on the Aristotelian picture, it is what exists and is good in and of itself. In regions where seasonal cyclones or periodic earthquakes are a fact of life, people probably think differently; the picture of a largely benevolent nature is something that they might have their doubts about. In recent decades, however, the seeds of the same doubts have been sown in more temperate latitudes. Should we drink water from this river or this or that underground source? Could these crops or that wild plant contain man-made toxins? There is virtually nothing that can be taken as given in our dealings with nature any more, and it sometimes seems frivolous to even approach the natural world without forearming ourselves with a mass of facts and figures about safety thresholds, use-by dates and the like. But do they really help when trust in its most basic form has evaporated? If the relevant toxin levels are shown to be below a certain safety threshold, does this mean that the meat in the freezer section of our supermarket is safe to eat? It doesn't – all it means is that the risk of exposing oneself to toxins by eating what reaches the supermarket shelves is considered socially acceptable.

A third traditional object of ontological trust is government, or, more generally, the state. Under premodern conditions, in Germany for example, until the end of the Second *Reich* in the early twentieth century, there was a general sense that authority was not manifestly unjust, that the police acted in accordance with the law, and that the laws themselves were relatively fair and reasonable. For someone like Hegel, the state was the highest expression of humans' substantial ethical being, *die substantielle Sittlichkeit*. Hegel's was a maximal concept of the state as an object of trust – the entire Hegelian sphere of political life was an institutional embodiment of ethics, a concrete instantiation of general principle. Now this sort of trust in the state as a *moral* authority has been deeply shaken by the events of the twentieth century. The fact that a state based on law and order can be deeply unjust, that police forces can be thoroughly corrupt, that agencies of the state can exercise their power arbitrarily and with consummate cruelty, that the laws of the state can involve flagrant contraventions of human rights – such things are deeply engraved in historical memory. True, we can point to progress: the paternal state of old has largely come to limit its role to administration. State power has been curbed with the help of a system of checks and balances. The activities of state agencies are now subject to judicial review in accordance with constitutional norms. But are such moves any help to us when trust in its most basic form has evaporated?

A fourth example is the economy, or, more precisely, the uneasy relationship between economy and society. As a rule, economic life is based on trust in numerous ways – trust that the money in circulation has the value ascribed to it, trust that prices are a reflection of the value of the goods that they are attached to, trust that prices will remain stable beyond the short term, etc. That may all sound trivial, though if it does that might well be because large parts of the Western world have been spoiled by a long period of reasonably low inflation and currency stability. In some places (Germany, again, for example), the shocks that were once sent through the system by runaway inflation and currency reform still resonate to a degree. However, turbulence in the stock markets is liable to spark panic in many quarters of society just about anywhere in the world. Although trust successfully sustains the overall economic edifice from one day to the next, the major disturbances caused by periodic crises demonstrate all too well that the entire world of economic cause and effect has something plainly incredible about it: our sense that the whole fantastic structure could collapse like a house of cards is palpable.[49]

My fifth example of ontological trust relates to other human beings – trust in the people one meets in the course of daily life. It goes without saying that in modern societies trust in other people can no longer be based on personal familiarity, or even personal acquaintance. Giddens is right to point out that most of the people we interact with on a daily basis are strangers. In Giddens' terms that is simply another consequence of social disembeddedness, or, as we might also put it, an aspect of the strangely contextless character of contemporary existence. Nonetheless, it is normal to presume that the people that we cross paths with are, generally speaking, well intentioned, polite, honest and sincere, that few are liable to flatly deceive us or do us violence. Yet this trust – without which the public sphere would fast turn into a rather nasty social jungle – looks more and more fragile: however exaggerated one might personally find the increased levels of theft or violent crime reported by sections of the media, the reports themselves alter the basic tone of social life; we note that our trust in our fellow human beings was predicated on the existence of a relatively civilized, peaceful society. The assumption we now realize we have long made has begun to fray: people are less inclined to let their children go about unaccompanied in public, they make sure that they lock their bikes with sufficiently intimidating bike locks, they carry as little cash on them as possible, they try to behave as inconspicuously as possible; those who are especially vulnerable take to carrying personal alarms or still more drastic forms of personal protection. It might be tempting to think that trust is greater in situations where one can rely on a degree of personal intimacy, for instance within the immediate circle of family and relatives. Yet the general outlook here is bad too. Whether relatives will stand by one in case of need, whether relationships within the family beyond those between parents will remain asexual, whether personal relationships need to be backed up by explicit legal contracts – all begin to look like open questions. The signs are unmistakable that what we once took for granted is breaking down: moves to introduce standardized rates of pay for house-work, the need for specific legal statutes criminalizing rape within marriage and the widely reported mistreatment of children all seem to point in the same direction.

As a final example of ontological trust, let us take trust in technology, our implicit faith that technical equipment is reliable, effective and safe to use, that technical facilities and institutions of various kinds will not fail us from one moment to the next. Such trust in technology is absolutely fundamental if everyday life is to run its course with relatively little friction in any society in which people constantly make use of devices whose workings are essentially mysterious to them. For the (literal and figurative) machinery of life to function, we have to be willing to make use of it without hesitation, without constantly assessing the potential risks and without taking safety precautions at every turn. Yet here too things look increasingly precarious; the fragility of trust in technology is amply demonstrated in the multiple safeguards that individuals and public authorities insist on, in the vast number of standards and specifications that technical devices have to meet, and in the threat of compensation claims when technical devices fail to live up to expectations.

Why, though, label all these various types of trust as *ontological* trust instead of adapting the notion of primary trust? Because, though trust of any form can indeed be nourished from personal sources, none of the five forms of ontological trust have their source simply in individuals' life histories. I have coined the term 'ontological trust' because this form of trust has its roots in the presumption that things, people and institutions are good *in themselves*. We could say that ontological trust is

encapsulated in the definitive conceptual figure of the Platonic theory of being –
an equation of being and goodness: the idea that *what something is* immediately
implies that it is *good*. Perhaps that seems too abstract a way of glossing our rough
conceptual position in this context, so let me put the thought more traditionalistically:
in proverbial terms, we might say that ontological trust is founded on the notion that
the earth is a benevolent mother (*mother* earth), that the state is a *paternal* authority,
a businessman an *honest* burgher, a worker an *honest* tradesman, a woman a *devoted*
housewife, and a man an *upstanding* member of the community.

The fact that some sort of residual ontological trust is still an active part of our
lives becomes palpable in the surprise that we feel when events take an unexpected
turn: when, for example, we hear of money losing its value in sudden economic
crises, police involvement in drug trafficking or the sexual abuse of children by
parents. When we make explicit what ontological trust involves, viz. the blind
belief that something or someone's existence is a guarantee of their goodness, it
can sound strangely naïve. To see this strange faith in operation, however, we don't
necessarily need to consider cases where our expectations are disappointed. We
need only examine the way the modern mind demands security – precisely where
premodernity was inclined to act wholeheartedly on trust.

The modern demand for security

Our growing loss of ontological trust undeniably has to do with the disembeddedness
that is one of modernity's most basic traits. Or, as we have also put it, it has to do
with the fact that the preconditions of human existence – what we all have to rely
on throughout our daily lives – are no longer something we become familiar with
through first-hand experience or personal knowledge. Yet the loss of ontological trust
surely also has to do with the fact that in modernity what things and human beings
are taken to be is no longer much of an indication of *what they are in themselves*.
Careers increasingly involve playing social roles or simply doing a job, mechanically
performing a function; there are few people of whom it could be said that work
defines their inner being. In the course of life, we simply have to reckon with the fact
that each and every person we meet is always going to be quite different to what he
appears to be. Nor is that analysis limited to human beings. Because what nature is
– and indeed exactly how much of our own being we have to regard as natural – is
defined by society, nature itself comes to appear less as the sustaining ground on which
we cannot but stand, and instead as a residual risk factor lying beyond our control
that we are compelled to accept. In addition to this, of course, comes a mounting
sum of personal and historical disappointments (everything from illness to natural
disasters). Yet do disappointments necessarily have to lead to the disintegration of
trust? They can also change the nature of trust, or lead to the development of trust as

an individual virtue. Either of those two paths are open to modern human beings: for example, when scientific developments force us to decide how much of our physical being we are willing to submit to technological manipulation and hence how much of ourselves we are willing to acknowledge as a natural given, we are simultaneously deciding to trust the nature whose limits we thus define. What trust means in this instance is yielding to *the nature that we ourselves are* as a spontaneous power active within us. Indeed, maintaining a sense of human dignity in today's world could well depend vitally on this type of trust – that is, on trust being practised as a virtue.

Something similar holds good for personal relationships. Here, too, the habit of trust is something we can learn to practise as an individual virtue. In relationships, trust means always giving the other the benefit of the doubt and, above all, a willingness to open oneself to the possibility of being disappointed by one's partner, friend, lover, etc. It also implies relinquishing a degree of control and avoiding mistrustful probing of others' actions and motives; in fact, it demands that one constantly takes for granted that the other means what he says. Here, too, we can say that the virtue of trust is the foundation of maintaining one's dignity and entering into dignified relationships with other people. Yet precisely because it acknowledges the other in his spontaneous subjective being, trust in one's dealings with others implies consciously giving up a degree of security. And that, as modernity continues to show us, is not everyone's cup of tea.

If anything, the opposite is true – the basic tendency within modernity that we confront here is a demand for *heightened security*. Philosophers might be tempted to trace the development back to Descartes' powerful experience of doubt and his quest for an unshakable foundation of knowledge. That would be a mistake though, for although early modern philosophy up until the age of German idealism might have sought and found certainty in the subject (in the self-experience of the individual human being), the path we have taken since the middle of the nineteenth century has, in many ways, led us in the opposite direction. For later modernity, it seems true to say that the individual has looked to the wider social realm *beyond* the domain of the self and its accomplishments for a deeper sense of assurance. Faltering ontological trust is not propped up by practising trust as a virtue, but by safety nets and security measures. Where we are not willing to grant nature our trust – whether it be external nature or our own human nature – we demand that it be managed. The feeling of being on intimate terms with ourselves, the positive bodily sensation of having something of nature in ourselves, has given way to suspicious self-examination. Essentially, we no longer trust our bodies. As a result, they have to be kept under constant control, every minor fluctuation has to be scrutinized, diagnosed; drugs ensure that the functioning of every part of our bodies is within a normal range, or, for those who demand more of themselves, within a range appropriate to peak performance. Nor do we have much faith in the wider natural world; the trust that nature will provide us with what we need, maternally and of its own accord, seems to be a thing of the past. Nature's exhausted body has to be constantly fertilized in order to maintain production levels; the end products of the process have to be tested for potential toxins and kept within chemically defined limits. Likewise, the behaviour of other human beings has to be monitored with as much technical ingenuity as we can muster, their performance levels carefully managed; as parties to social transactions of any kind, their behaviour has to conform to a strict regime of guarantees, obligations and enforceable legal provisions.

Nor do we feel as if we can rely on the general pool of technical know-how any more than we can on social behaviour. Products of all kinds have to be produced according to strict norms, incorporating wide margins for error and back-up systems for emergencies. No technical device can be built in such a way that the dimmest user could ever get into difficulty using it. With everything so well regulated that engineers are left almost no room for innovation, some might conclude – perhaps even with a certain degree of satisfaction – that our trust in technological products is finally rational and justified.[50] They would be overlooking the main point, however, which is that the whole manifold apparatus of control – all the boundary conditions, guarantees, regulations, technical norms, declarations of liability, safety measures and security checks – has been introduced precisely to counteract the disintegration of trust.

At bottom, what this interplay between the loss of trust and the demand for security generates is an escalation of the problem. The loss of trust so easily starts off as little more than a niggling doubt, but in the instant we decide to remedy our doubt by introducing some sort of security measure, we forfeit a fraction of our readiness to accept uncertainty of any sort. Having reached a certain level of security, every accident, every human error, every technical malfunction leads to calls for tighter security, new standards, new technical guidelines: security is boosted and the basis of trust becomes ever more tenuous. Over a century ago, Max Weber argued, as we have seen, that the advance of rationalization, which he held to be deeply characteristic of modernity, has confined human beings to life in an iron cage. Perhaps the mechanism that drives our loss of trust and our co-ordinate demand for security is the central cause of our predicament on this score.

In any case, the process of escalation leads to so much further standardization of behaviour and so much further legal regulation of production that the individual layman can hardly ever cope without expert help, or else ends up in something like the position of a savage within civilization should he choose to simply go out and buy something, hammer a nail into a wall or plant his garden however he wishes, without considering what precautions might need to be taken. The security systems we have built up around ourselves have grown so complicated that they themselves condemn the individual to insecurity, or at least to *feelings* of insecurity. What can we do but take refuge in trust of a second – or perhaps a third – order, i.e. in conventional faith in experts? And what does doing so amount to in the final analysis, but resigning ourselves to fate? The specialists will have the problem well in hand, or so we think; if margins for error have been allowed for, if a product's key specifications are within stipulated limits, then it can do us no harm, or so we assume. And, indeed, there is no denying that such a conventional sense of trust is enough to keep daily life in the modern world from coming off the rails. Yet was the project of contemporary modernity not launched precisely in order to make good our loss of trust by introducing more security measures? Obviously, we only succeeded in making good the loss by taking temporary refuge in a higher level of trust – in second-order trust, though the latter is much more fragile than trust in its ontological form.

The first signs of the erosion of second-order trust are well and truly visible. Thus, the general public's once nigh-on universal esteem for the medical profession has been tarnished by the proliferation of iatrogenic illness and by the ruthless commercialism with which medicine has often been practised. Thus, the legal layman who finds himself in court often has the feeling that the proceedings have little to do with his case – that, in practice, the law is a game that lawyers play among themselves. Thus, the exploits of the scientific community – and in particular its claims about the effects of scientific breakthroughs on daily life – are met with an increasing degree of scepticism. Thus, the use of chemicals in all areas of agriculture and food production is seen, almost automatically, by a growing number of people as something dubious if not downright dangerous. Under the circumstances, we might indeed ask – what forms of security would we have to dream up in order to restabilize a modernity that seems to perpetually endanger itself and its own existential basis should our deepening loss of trust in experts become a genuinely mass phenomenon?

Free scientific enquiry and its limits

Free scientific enquiry as a basic right

The freedom of scientific enquiry is enshrined in the Basic Law of the Federal Republic of Germany as a fundamental right. Paragraph 3 Article 5 states that: 'Arts and sciences, research and teaching shall be free. The freedom of teaching shall not release any person from allegiance to the constitution'.

The remainder of Article 5 deals with freedom of opinion and freedom of press. The fact that the right of scientific enquiry appears in the same context as freedom of the press is both a good indication of its historical derivation and a pointer to how it should be interpreted. Freedom of scientific enquiry is evidently not to be taken as a mere aspect of freedom of opinion or freedom of the press. (The Austrian constitution, for instance, clearly treats the two as *separate* fundamental rights.) The historical source of the article lies in the Basic Law of 21 December 1867 and its validity is based on the Constitution of 10 November 1920. Article 13 of the latter document reads as follows:

> Everyone has the right within the limits of the law freely to express his opinion by word of mouth and in writing, print or pictorial representation. The press may be neither subjected to censorship nor restricted by the licensing system. Administrative postal distribution vetoes do not apply to inland publication.

Freedom of scientific enquiry has been clearly set apart. It receives separate mention, but not until Article 17: 'Knowledge [*Wissenschaft*] and its teaching are free'.

What follows in the remainder of Article 17 of the 1867 and 1920 constitutions are further specifics relating to education. Though the issue is set out differently in the Austrian constitution of 1920, this need not imply that freedom of scientific enquiry was actually *understood* any differently in post-War Austria.

The key point for us is that the Basic Law of contemporary Germany accords free scientific enquiry the status of an elementary constitutional right: from the point of view of the Basic Law, scientific freedom must be seen in practical terms as legally immutable. That brings with it certain problems – problems indeed that have already arisen in recent decades in relation to the so-called right of asylum, Article 16. Here, the constitution states baldly: 'Persons persecuted on political grounds shall have the right of asylum'.

Such a far-reaching right of asylum – whose origins lie in the self-image of post-war Germany as a nation owing its existence to the defeat of National Socialism – could only be restricted through the addition of supplementary clauses specifically narrowing the spectrum of those entitled to asylum. As the Basic Law currently stands, the immutability of the basic right to scientific freedom is underscored by the fact that it applies without any formal restriction. (A comparable case would be the right to life, which is subject to the proviso set out in Article 2 Paragraph 2: 'These rights may be interfered with only pursuant to the law'.)

In essence, the Basic Law interprets freedom of scientific enquiry as something *so* fundamental that it makes no provisions for limiting it by means of specific laws. In the present essay, we want to moot the case for modifying the principle of free scientific enquiry. The question we want to ask is this: is there cause for limiting the scope of the principle as it stands, viz. without qualification? Our answer will be in the affirmative.

Why should the constitutional right to scientific freedom be changed?

In the Basic Law as it currently stands, freedom of scientific enquiry belongs in a wider frame of reference which also includes freedom of the press and freedom of opinion. For states like the German Federal Republic and the Republic of Austria, the latter freedoms are, undoubtedly, of fundamental significance, for both are states whose foundations are interpreted as *liberal* and post-absolutist, i.e. states whose self-conception has been shaped, and shaped fundamentally, by a rejection of feudal and absolutist forms of authority. Freedom of opinion, freedom of the press and the abolition of censorship were, in other words, basic to the way these states came into existence in the early twentieth century and few would dispute that the importance of such a constellation of basic rights was appreciably heightened by their abuse, first during the Nazi period and subsequently during the years of communist rule in East Germany.

For democratic states based on the rule of law, the existence of a liberal public sphere is crucial, for it is here that the *volonté generale* takes shape, here that public criticism can be voiced. One need look no further than Kant's essay 'What is Enlightenment?' for a formulation of the idea that is second to none in clarity and depth. For Kant, the cornerstone of Enlightenment is the so-called 'freedom of the pen', the possibility of presenting one's unadulterated opinion before the general public (though to judge by his willingness to restrict the freedom of the pen to the circle of the learned, Kant does not quite escape the subservient spirit his essay otherwise so eloquently casts off):

> The public use of one's reason must always be free, and it alone can bring about enlightenment among mankind; the private use of reason may, however, often be very narrowly restricted, without otherwise hindering the progress of enlightenment. By the public use of one's own reason I understand the use that anyone as a scholar makes of reason before the entire literate world.[51]

Now the fact that the principle of scientific freedom enshrined in the current Basic Law has its roots in the Enlightenment and, in particular, in the battle to establish a liberal public sphere free of censorship indicates clearly that the Basic Law interprets science as a quintessential vehicle of Enlightenment. Historically, this makes sense. Science was, indeed, one of the means by which European society aimed at, and succeeded in, liberating itself from traditional authority and traditional world views, in particular the authority and world view of the church; well into the nineteenth century, the social significance of science was seen to lie in its basic conception of reality, and in its ability to challenge existing conceptions. Yet the last historical episode in which it still played something like this part was the great culture war of the nineteenth century – the confrontation with the church over Darwinism. The latter half of the century saw a decisive shift in the cultural meaning of the scientific enterprise. In the age of high capitalism, science became a major force of production, starting with the manufacture of chemical dyes, then, soon after, the use of chemistry for a range of industrial purposes, from the production of fertilizers to poison gas. In later decades, the twentieth century was to give us systematic scientific pharmaceutical research, systematic military research and, in more recent times, the systematic development of biotechnology.

Science has become an economic and military factor second to none. Companies with a scientific edge give themselves a considerable competitive advantage, while nations with an edge are assured of military power and prestige. Since demonstrating its adaptability for production and warfare, scientific knowledge has thus been privatized to an ever-increasing degree. Patents and classification have been central to the cultivation of a certain ethos of research – research that is off limits to the general public or is only accessible with considerable difficulty. In practice, it is clear that science is simply no longer something with any *essential* reference to the general public or any *essential* grounding in the search for truth. While for the Enlightenment science was a project that was addressed to the general public, and while Enlightenment science viewed itself as *working to enlighten this very public*, science today has been largely functionalized (finalized, as one might put it using the language of the Starnberg circle[52]). What that means in effect is that science is practised for certain economically defined purposes. However things might stand in theory, science has become a particularistic and, in some circumstances, even a nationalistic enterprise – a situation that the authors of the Basic Law completely failed to register.

In short, the founding fathers of the post-war German state endorsed the principle of scientific freedom as it came down to them from the scientific self-image of the Enlightenment and the foundational documents of a liberal society, viz. from the various draft constitutions dating from the earlier nineteenth century.[53] Not only do they *not discuss* the fact that science has undergone sweeping changes since the mid-nineteenth century, above all in no longer taking its bearings from any sort of ideal of general enlightenment, they do not even appear to be *aware* of the historical shift. If they had been, one would at least have expected the principle of scientific freedom to have been treated separately from the issue of freedom of the press and freedom of opinion.

A second reason for altering the principle of scientific freedom is the fact that the current Basic Law interprets scientific freedom as a right conferred on *individual* researchers, even though in reality science has long been a form of *collective* activity.[54] As matters stand, scientific freedom belongs with the other constitutionally guaranteed liberal freedoms as a right of individuals to be upheld *against the state*. Again, we glimpse the heritage of the rights set out in the Basic Law and earlier liberal German constitutions – their origin in a political process of emancipation from the absolutist state of the eighteenth century and the totalitarian state of the twentieth century. Like Austrian constitutional law or indeed like the 1948 Universal Declaration of Human Rights, German Basic Law interprets freedom essentially as *freedom from interference*, viz. by the state, the church or any other major social power. Freedom of speech, freedom of conscience and creed, the inviolability of person and home, etc., are formulated as fundamental rights vis-à-vis such powers, in particular as fundamental rights of the individual vis-à-vis the wider community represented by the organized state.

Yet it is obvious that science has long ceased to be an affair of individual scientists; in by far the largest subsectors of scientific activity, in the natural sciences and in technological research and development, it has actually become *impossible* for individuals to carry out scientific research on their own. What the individual can do as a researcher – everything down to the specific topics he can choose to research – depends fundamentally on the institution where he works and on the technical facilities that are available. In general, there is no avoiding the fact that research is organized as a form of teamwork. As a result, scientific publications regularly bear the name of up to 20 separate authors. In a situation like this, the principle of unlimited scientific freedom – interpreted as freedom of the individual researcher – seems bizarrely out of tune with the reality of standard scientific practice.

Anyone who doubts that our problem has a practical political dimension should recall the controversy surrounding the Hamburg Institute of Technology in the early 1980s, where academic opposition to arms research led the governing council of the university to place a wholesale ban on research for military purposes. The decision was overruled by the then Hamburg State Minister for Science, Klaus Michael Meyer-Abich, on the grounds that it violated Article 5 Paragraph 3 of the Basic Law. Meyer-Abich's legal review of the case concluded that 'the Basic Law guarantees the Institute's staff the right to participate in defence research and projects with links to the military'. The wording of the ministerial veto shows clearly that Meyer-Abich had interpreted scientific freedom as an *individual* right of *individual* researchers. What he was saying in effect[55] is that rulings of institutions aimed at (e.g. ethically) limiting the choice of research topics are in contravention of the basic law of the state.[56]

The travails of the Hamburg Institute bring us to a third reason for rethinking the principle of free scientific enquiry. As we have seen, when the principle of scientific freedom was formulated in accordance with Enlightened ideals of basic human freedoms, participating in public intellectual discourse was of the essence of science. Kant, as we have seen, thought of scientific freedom as 'freedom of the pen'. In the 200 years since Kant, science has become something different. The heart of the scientific enterprise is now research, and research of a very particular kind; publishing one's findings in a broadly readable form or participating in public discourse are no longer considered to be actual scientific work, all the less so as science has been privatized; in some circumstances, commercial, national, legal or institutional considerations may well mitigate *against* publication. Yet if science is essentially research of this kind, should the principle of free scientific enquiry really be taken to mean that scientists have the freedom to research whatever they please for whatever motives or with whatever ends they please? Why should we regard research into any area whatsoever, or for any reason whatsoever, as necessarily legitimate from the point of view of wider society? The furore over arms research in Hamburg points to the sorts of competing perspectives that can exist on what is and is not a legitimate object of research. The history of science during World War II ought to have demonstrated well and truly, however, that there is such a thing as unethical scientific research.[57]

In more recent times, the need to set boundaries to the freedom of research has re-emerged in the field of biotechnology, and in particular in relation to the use of human cells and the human genotype as a whole. It is to this tangled knot of contemporary ethical, legal and scientific problems that we now need to turn.

Limiting the freedom of science qua freedom of research?

The formula set out in Article 5 Paragraph 3 of the Basic Law provides for a single kind of limitation to scientific freedom, viz. where research contravenes the researcher's higher-order allegiance to the constitution itself. Although the limitation relates chiefly to freedom of instruction, one can certainly argue that it ought to apply to science as well – the main problem being that no particular limitation to scientific research would seem to follow from the mere notion of allegiance to the constitution. Specifically, the problem is that the Basic Law sets no limits on freedom of research that relate to actual *objects* of research. When it comes to restrictions relating to researchers' purpose or intentions, the constitutional room for manoeuvre looks even smaller.

One possibility for restricting scientific freedom qua freedom of research, though a trivial one, seems to arise where there is a *conflict* between several of the rights set out in the Basic Law. This would be the case where research could be regarded as contrary to human dignity, for the Basic Law subordinates all other basic rights to the right to human dignity; the article which enshrines human dignity, Article 1, takes priority over the rest of the document in every sense of the word. (Most of the experiments on human beings performed during World War II would clearly be deemed contrary to the Basic Law on the grounds that they violate the principle of human dignity.)

Justifying limitations on research on the grounds that objects of research stand under the protection of the Basic Law is a more difficult matter than invoking the provisions of Article 1. Nevertheless, the former is the preferred strategy of those who nowadays aim to set limits to research into biological life, in particular to experiments using human genetic material such as stem cells. In line with Article 2 Paragraph 2, human life does indeed stand under the protection of the Basic Law. What is doubtful is whether Article 2 Paragraph 2 provides a solid enough rationale for limiting research on *any sort of human life*, as opposed to individual human beings. The wording of the law is all important: 'Everyone has the right to life and physical integrity'.

Although human life is an avowed object of legal protection, Article 2 Paragraph 2 specifies that 'human life' is to be interpreted as the *life of human beings*. 'Everyone' who has a 'right to life and physical integrity' clearly refers to every *person*, perhaps indeed every citizen. People, in short, are the bearers of the constitutional rights to life and freedom from bodily harm.

Now, if we ask whether research on human embryos or embryonic stem cells passes constitutional muster, it seems obvious that Article 2 Paragraph 2 stops short of ruling out such research. So much became clear from the arguments and the counter-arguments within the 1948 parliamentary committee responsible for the wording of the law in its current form. Since then, the Federal Constitutional Court has ruled that unborn life *does indeed* also stand under the protection of Article 2 Paragraph 2.[58] Nonetheless, it is easy to see why the exact wording of the article sparked broad public debate about *when* human life is to be regarded as the life of an individual human being. Opinions on the matter diverge widely: for instance, the orthodox Catholic view is that *conception* marks the point at which an individual human being comes into existence, while Jewish teaching dates the life of an individual person to *birth* (i.e. the separation of the infant from the maternal womb).

Since 1990, when the Embryo Protection Law was passed into effect, controversy has largely died down. The scientific use of human embryos has effectively been restricted to use for reproductive purposes; essentially, the law provides for the manipulation of embryonic life in assisting with conception, and for no other reason. Freedom of research is thus subject to qualification by means of a specific law that forms part of the penal code, though clearly a part that must be read as a gloss on the conception of human life contained in Article 2 Paragraph 2 of the Basic Law. Section 1 of Paragraph 2 of the Embryo Protection Law, entitled 'Misuse of Human Embryos', reads as follows:

> Anyone who sells a human embryo, whether it is artificially created or removed from the body before the conclusion of nidation in the womb, or who acquires, uses or turns over to others such an embryo for purposes other than preservation and care, will be punishable by fine or by imprisonment for a period of up to three years.

As we have indicated, its specific wording gives the Embryo Protection Law a legal pre-eminence that is quite disproportionate to its place within the scheme of punishable offences: human embryos are here accorded a legal dignity that agrees with the Kantian categorical imperative in its full force: they are never to be regarded *solely* as means, but as ends in themselves, as the Categorical Imperative famously demands all human beings be regarded: 'Act in such a way that you treat humanity, whether in your own person or in the person of any other, never merely as a means to an end, but always at the same time as an end'.

Clearly, it would be untenable to argue on these sorts of ethical-legal grounds that the human embryo is the bearer of a human dignity separate from the dignity of the fully formed individual human person; the wording of the Embryo Protection Law can thus be taken to mean that a human embryo is to be understood as an incipient form of personhood, and it is as such that Article 1 of the Basic Law describes the embryo as the bearer of a right to life and physical integrity.

The matter is far from theoretically straightforward and, before concluding, it seems in order to briefly review two very different approaches to these matters. In the period since the law came into force, for instance, Habermas has been thinking along quite different lines to those that implicitly underpin the law as it stands. In line with his view of the ethical centrality of reciprocal relations between people, in *The Future of Human Nature* (2001) Habermas proposes a distinction between *human dignity* and the dignity of *human life*.[59] 'Human dignity in the strictly moral and legal sense of the term', he writes, 'depends on an essential symmetry between the parties to a relationship. Human dignity is not a natural attribute like intelligence or blue eyes that one "possesses"; the notion is rather the marker of a certain "inviolability" that can only have a meaning in interpersonal relations of reciprocal recognition and in egalitarian forms of interaction'.[60]

Habermas' view contrasts sharply with the type of morality of responsibility to be found in the work of Hans Jonas, for according to Jonas, morality has its primary locus *precisely where relationships are marked by asymmetry* – in our responsibility for children, for 'nascent forms of life', even for future generations, for obvious reasons also in our responsibility for the handicapped. For Jonas, morality makes its claim on us precisely in those situations where there can be no question of reciprocal recognition. If we can speak at all of our moral responsibilities vis-à-vis incipient forms of human life, then it has to be for this very reason, viz. because reciprocal recognition is *impossible*.

The further one pursues Jonas' line of thought, the more it must seem that Habermas' distinction between human dignity and the dignity of human life destabilizes the intellectual ground on which the Embryo Protection Law is built. Habermas' surprising turn towards a notion of human nature in recent times is, in this sense, anything but a step beyond discourse ethics: in spite of his professed preference for a 'species ethics', it would seem that the author of *The Future of Human Nature* is yet to seriously contemplate the possibility that *the nature we ourselves are* is an essential facet of what it is to be human.

The point comes out most clearly in the above excerpt of the text in Habermas' rejection of the notion that human dignity is a natural endowment. The 1948 Declaration of Human Rights, it must be said, takes its stand on a very different interpretation of the situation: human rights are ascribed to human beings as members of the human family in the preamble to the 1948 text, and then even more explicitly in Article 1, on the grounds that the individual is *born human*: 'All human beings are born free and equal in dignity and rights'. Human dignity on this reading is not something that human beings *acquire*; it is something that is imputed to them because they are, by nature, members of the human species. And if human dignity is to remain independent of particular qualities, abilities or achievements, then that is something all important, particularly if the risk of denying dignity to the mentally handicapped is to be systematically averted.[61]

In short, since dignity is something we ascribe to individual human beings as part of a natural biological heritage, that natural heritage should be guarded against manipulation. Human life, as the source of the life of individual human beings, should be regarded *ab initio* as having inalienable dignity.

In Germany today, the Embryo Protection Law places a clearly palpable limit on scientific freedom because of the restriction it places on what can be considered a legitimate object of research. The field of social tension that surrounds the law is an extraordinarily powerful one for international competition between stem cell researchers, as well as economic expectations born of naïve faith in unfettered technological development, exert pressure that clearly runs contrary to the intentions of those who framed the law. This is why we would argue that the situation as it stands remains problematic; the fact that the principle of scientific freedom goes unqualified in the Basic Law and hence can only be indirectly and arduously subject to restriction where conflict between basic rights arises, or else where there is room to be flexible about what stands under the protection of the Basic Law (in this case – human life), encourages various parties to the debate to play fast and loose with the penal code where their own interests are at stake.

The welcome but problematic limitations to freedom of research that exist in the form of special laws like the Embryo Protection Law are under threat from various quarters. The theoretical threat becomes clear in Habermas. In *The Future of Human Nature*, we read for instance that 'legal concepts such as human rights and human dignity do not just forfeit their conceptual sharpness when they are overextended in counterintuitive directions; they also lose their critical potential'.[62] However, the tenuousness of the present state of play is also evident from a legal point of view in the efforts of scientific lobby groups, who, with the backing of the medical industry, have successfully introduced legal limitations of their own, viz. to the Embryo Protection Law. The so-called Stem Cell Law of 28 July 2002 brings out the conflict forcefully in Paragraph 1:

> Purpose of the present law, 1: the purpose of the law is to guarantee freedom of research in light of the state's duty to respect and protect human dignity and the right to life . . . 3: to specify conditions under which the importation and use of embryonic stem cells is permitted for exceptional purposes of scientific research.

In effect, the law opens the way for the very embryonic stem cell research that was barred by the Embryo Protection Law; the Stem Cell Law clearly envisages the importation of stem cells from countries where stem cell research is less tightly restricted. One indeed has the impression at times that in the two centuries since its formulation during the Enlightenment, the principle of scientific freedom – which in its original form can hardly have implied absolute freedom of research, let alone the freedom of individuals to experiment on whatever they please for whatever purposes they please – has turned into a powerful instrument in the service of particular social groups and vested economic interests. Unless its misuse by the medical–industrial complex is contested, the status of scientific freedom as a genuine ethical-legal precept can only become more tenuous.

Borderline situations
in technological civilization

We owe the concept of borderline situations to the philosopher Karl Jaspers. In its original context, it is part of the project of *Existenzerhellung*[63] – what we could call Jaspers' philosophical anthropology. *Existenzerhellung* involves an attempt to present human life from a particular evaluative standpoint: the standpoint that one's life is something one must take *responsibility* for. In the analysis of what this involves, the notion of borderline situations plays a decisive role.

What are borderline situations? First and foremost, they are situations that bring out the conditions – the very boundaries or limits – of human existence. Jaspers defines a situation as 'whatever presents itself as real to a human subject interested in reality as an existing being'. A situation, in short, is *not* in any simple sense a state of affairs or a neutral constellation of objects, but a state of affairs or a constellation of objects in their essential relevance for an individual human being.[64] Jaspers goes on to define borderline situations as situations that are ineluctably part of human existence: 'I find myself in the situation that I am always in situations, that my life is impossible without suffering and some degree of struggle, that I unavoidably incur a degree of guilt, that I am mortal – these are what I call borderline situations'. To begin with, that may seem a surprisingly open-ended pronouncement in the sense that it altogether fails to distinguish borderline situations in terms of *particular* situations – something most evident in Jaspers' very first example of a borderline situation, the situation that I am always in situations. The intention is obviously to include everything within the scope of the term 'borderline situation' that is *necessarily* and *unavoidably* part of human life. However, the concept of a borderline situation immediately comes to have *critical* meaning if we take just one more step – perhaps a step beyond Jaspers – by noting that in the *normal* course of life we don't notice the borderlines of human existence. It is in situations involving danger, or, alternatively, in situations involving a heightened sense of the pleasure of life, when existential borderlines perhaps first become noticeable and when we thus begin to notice some of the peculiar depths of human existence. Jaspers' own view is simply that we take part in human existence *fully* (for Jaspers that means *consciously*) when the borderlines of existence become apparent. His argument is that placing ourselves in borderline situations, seizing the opportunities they present, in a sense represents the crown of human existence (what he terms *Existenz* in contradistinction to life in general): 'We become who we are by entering into borderline situations with our eyes open. Experiencing borderline situations and existence [*Existenz*] are one and the same'.

So let us try to adapt Jaspers' concept of borderline situations to the present day. Our average everyday mode of existence is shaped by the fact that we live in a technological civilization. And life in a technological civilization, let us say, is distinguished by three main features.

The first feature is the dominance of technoscientific knowledge. We might also say that we live in a *knowledge society*, which in more precise terms means that technoscientific knowledge has come to subordinate or suppress all other forms of knowledge and has consigned education in other forms of knowledge to a marginal existence. The prime consequence of this technification of knowledge is that individual human beings are no longer capable of making autonomous decisions either about matters of vital importance to them or about smaller questions. Whether it is matters of life and death or a matter of feeling marginally perkier or happier, modern human beings surrender the direction of their lives to experts, be it directly or by consulting the vast range of books proffering expert advice.

The second feature of life in a technological civilization is the delegation of work, movement and thought itself to machines and technological systems. Moreover, the various types of communication that make up such an essential part of human life become more and more technically mediated; perception, speech, writing, learning and much else take place to an ever greater degree in technological media, or through what we collectively call *the media*. As a result, our bodies become superfluous for the purposes of most social existence and for many of the actions and transactions of life: step by step, what we seem to be slipping into is a primarily *non-corporeal* form of existence.

The third feature of a technological civilization is the averaging out of human life stories via any number of forms of insurance, by the agencies of the welfare state, by medically stabilizing and technically standardizing human beings' physical condition and by normalizing expectations about the level at which human bodies should perform. The prime result of this neutralization of life's contingencies is the curiously *eventless* character of contemporary human life. Another is the way we live out our emotional lives largely in imaginary or fictional spaces (film, television and the internet) to make good the emotional thinness of our experiences of material reality.

With these characteristics of existence within a technological civilization set out in rough form, we can now ask where they can be situated in relation to existential borderlines in Jaspers' sense: how do they relate to the possibilities of experiencing – and experiencing consciously and actively – the depths of human existence? The answer also comes in three parts.

First, note again that borderline situations are something we *inevitably* have to confront. As Jaspers defines them, they represent precisely those conditions of human existence that are impossible to avoid. As we have seen, there is no changing the fact that we always find ourselves in situations, that we are mortal, that we experience pain, that life involves incurring guilt and being a party to conflict. Yet what is deeply characteristic of our contemporary form of existence within a technological civilization is that these borderline situations are increasingly concealed from view, first and foremost through what we have been calling technification – the delegation of life's problems to experts, the neutralizing of life's fluctuations through insurance and, above all, the use of technical devices for medically manipulating the human body, the use of drugs to alleviate the pain of childbirth or the relegation of death to the intensive care units of hospitals. The dominant mode of thought plainly suggests that death, guilt and suffering are, in principle, avoidable and, in practice, something to screen oneself from.

Thus, if borderline situations are, in fact, unavoidable, they are bound to take the individual by surprise. Given the guilt-averse, suffering-averse and death-denying mechanisms of a technological civilization, we are bound to experience existential borderlines as a rent in the fabric of conventional life. The concept of borderline situations needs to be updated accordingly: the borderline situations of today are situations in which the safeguards of habit and the horizons of meaning provided by our sense of normality fall away. They are situations marked by discontinuity and extremity.

However, precisely because the experience of existential borderlines belongs outside the usual run of life in technological civilization, borderline experiences also become something that individuals consciously seek out or intentionally try to provoke. Moreover, the techniques that are available for this very purpose – what we might call, with a generous dose of irony, provocative forms of behaviour – are of various kinds and are not always easy to recognize as such: everything from ascetic self-control and deliberate fasting to extreme sport and extreme tourism, all the way to programmatic drug use and subcultural cults of violence. The whole field is fraught with questions: the urge to seek out and intentionally enact borderline situations appears suspiciously like the product of *ennui*, it can smack of sensationalism or plain attention seeking, and it easily exposes the provocateur to the charge of irresponsibility. Nonetheless, there is no ignoring either the fact that provoking borderline situations is a characteristic part of life within a technological civilization or the fact that such provocations express a genuine human need – in basic terms, the need to experience what it really means to exist as a human being. Borderline experiences are experiences that human beings are deprived of by the normal run of life in a technological civilization. Little wonder then that they seek them out in the interstices of civilized life, or, wittingly or unwittingly, make use of technical means in doing so.

So, we have our three key features of contemporary borderline situations: concealment, surprise and provocation are the three things that mark out the experience (or absence of experience) of borderline situations in a technological civilization. Individually and together, they challenge us to ethical reflection. They pose the question whether the concealment of borderline situations, as it systematically comes to pass, is something we should always go along with: what stand should we take vis-à-vis the provocations of those who are consciously seeking out existential borderlines? And they raise the further question of how to go about consciously preparing ourselves to be surprised by borderline situations.

Here we notice immediately that there was once an ethics – perhaps we should say an ethical standpoint, or simply an ethos – whose prime concern was actually acquiring practice in dealing with borderline situations. Though admittedly it is an ethic that many would repudiate today as a cult of manliness and war, the principal imperatives of this ethics were resisting temptation, showing courage, tolerating pain and looking death in the eye. (*Virtus*, the Latin word for virtue, originally meant manliness and nothing more.)

But hasn't this kind of ethic of manly fortitude been rendered superfluous by the way we live today? And should we not be glad to live under the peaceable conditions that life within a technological civilization generally brings with it? Perhaps not entirely – for what ethics *is* today (in point of fact) is behaving according to conventions (doing what we are expected to do, be it within the family, at school, within a peer group, as the employee of a company or a citizen of a state), whereas borderline situations occur on the margins of conventional life, precisely where conventions no longer suffice and where life becomes – surprisingly – serious.

If there are any general recommendations that follow from what we have said, then the first would have to be to resolutely draw back the conventional veils that hide borderline situations from us. As Jaspers says, borderline situations are something we have to confront at some stage – indeed, avoiding them means being deprived of an essential experience, or rather *depriving ourselves* of an essential experience. Here, again, it is important not to let all of our needs be supplied by teams of experts, to exercise an active form of informed consent as patients, should we become ill, and to ensure we will be able to die a dignified death. Equally, we should not try to insure ourselves against every one of life's possible exigencies. Bearing part of the risk of one's existence oneself is an essential precondition of leading a responsible life.

So much for the problem of exposing concealed existential borderlines. What more can be said about provoking experiences at or near the existential borderline? How to classify the activities and lifestyles of the provocateurs of existence? We have already argued that these activities and lifestyles are justifiable up to a point, or at least understandable in the context of contemporary life. Nonetheless, what we are dealing with here are forms of life that bespeak a certain despair – outbursts that can never be brought into harmony with the regular demands of life and sometimes seem like straightforward acts of spite against any sort of regularity. In this case it seems to me, we would do well to ask whether such existential provocations could be given a moderate form, whether they could be used as a way of acquiring practise in borderline situations, or in preparing us for the moments when life gets serious in the full existential sense of the word. Ascetic self-control can teach one something about one's dependence on nature, as can self-imposed fasting. Sport can inure one to pain and lead to the practice of relatively simple forms of self-overcoming. The experience of extremity that drugs are meant to provoke can also be attained through meditation: testing the limits of the human could take in forms as various and as sophisticated as human beings are capable of imagining. And finally, what about violence? When Jaspers makes the point that conflict belongs ineluctably to human life, what should that be taken to imply? In our view, practised self-assertion and a readiness to enter into confrontation, at least in certain circumstances, should form an indispensible part of human beings' preparations for dealing with borderline situations. Instead of overt cultures of violence – let alone the stylized thuggery of Hollywood movie scenarios – the arts of self-defence (Japanese war-games, etc.) seem like culturally far more promising starting points.

Adapted as part of new forms of practice, the strategic seeking out of borderline situations involved in the more extreme forms of existential provocation could serve to prepare individual human beings for the borderline situations that life has in store for them. Such practice would prepare us for those junctures where life takes on a depth of seriousness, viz. when we come up unexpectedly against borderline situations – as we inevitably will.

3

Understanding technology: Use and entertainment

Technical gadgetry

What's the use?

We can start by confirming a prejudice that we all share: technology is useful. The idea runs so deep and seems so obvious: we think we can only understand technology in terms of its end – the purpose it serves. Whether we are talking about a simple tool, a machine or a whole technical system, if we want to understand what we are dealing with, we have to infer it from the function of the gadget or the system that we have in front of us. New concepts of technology – the idea of technology as a medium, as infrastructure, or as a material *dispositif* – have done little to change the way we think on this score. As much as we struggle to get beyond the idea of technology as a means to an end, it has a way of creeping back into our thinking, time and again. To get a sense of other possibilities, we seem to have to turn away from the world of material reality to the world of art.[1]

The tight-knit conceptual connection between technology and use, however, is not so much a product of the actual history of technology, as a product of the *theory* of technology; it may well be the abiding influence of Karl Marx, who thought of technology as closely tied to the concept of use. Prior to Marx, there had indeed been books about technology and there was no absence of interpretation and reflection. Marx, however, was probably the first to develop a full-blown theory of technology, and certainly the first to see technology as a *social process*. Importantly, his theory of technology came into being during the time of the Industrial Revolution – i.e. in the first phase of history in which technological development went hand in hand with deep and radical social change. It is understandable that Marx thus defines technology in terms of forces of production and thinks of it primarily in relation to labour – labour considered namely as the appropriation of nature. Technology, for Marx, is a means with which people work, something that increases the efficiency of production.

Where does this picture originate? What Marx observes taking shape in his own time is a development that was changing the form of work and hence changing the relations of production. If, for centuries, technology in the form of basic hand tools had been a means for bringing human skill and labour power to bear on the appropriation of nature, the Industrial Revolution was to bring about two radical new developments. Marx explains them with reference to two emblematic new inventions: the spinning machine symbolises the mechanization of manual labour, while the steam engine symbolises the replacement of human beings by machines as a source of raw physical power. The development as a whole was to revolutionize relations of production. Though workers – for example in the publishing and manufacturing industries – might formerly have been dependent on the owners of capital, as a rule they still owned their own tools, and the surplus value of production was a direct result of the way they transformed the raw materials by means of labour. With the advent of the Industrial Revolution, the relations of production were to change: workers were no longer the owners of the means of production, and surplus value came to be generated largely through the use of machines. We can still talk about what was going on in terms of an increase in the productivity of labour; yet because that increase came about through the use of capital in the form of machines, the individual capitalist, who owned the machines, claimed the right to appropriate the surplus value for himself. Technology – in this case the technology of mass production – thus came to produce and reproduce a certain sort of class relation between labour and capital.[2]

The general scheme of Marx's interpretation of technology has defined mainstream theories of technology ever since. Technology, on this picture, belongs in the realm of *necessity*, *not* in the realm of freedom. It is a means by which human beings appropriate the bounty of nature it helps to provide them with a living, or else it helps them to improve their living conditions. Technology, in short, is there to serve human purposes, and ultimately the purpose it serves is reproducing the human species.

Yet constructing that persuasive frame of reference is still not the full extent of Marx's influence. Marx's interpretation of technology has also been academically institutionalized in the sense that the history of technology is generally considered part of economic history and not, say, part of the history of art. Very different theories of technology – theories which are certainly far from Marxist in their political orientation – share with Marx's theory the view of technology as a utilitarian institution, as a means or a set of means for useful ends. The first modern philosophy of technology, Ernst Kapp's, is in this category because, as we have seen, Kapp interprets technology as an extension of and a substitute for the organs of the human body. In Kapp's anthropological theory of technology, interpretation starts out from the organic functions of the human body. According to Kapp, technology supports human beings' organic functions by increasing their effectiveness and scope, and finally by taking over their role completely. As in Marx, on Kapp's interpretation, technology serves the purpose of survival and, ultimately, the reproduction of the species.

Arnold Gehlen's theory of technology, which is explicitly conceived as an anthropology, runs along similar lines.[3] Gehlen, as we have also noted, takes up the idea traceable to the Greek sophist Protagoras that human beings are, by nature, deficient beings.[4] Having been insufficiently endowed by nature, humanity creates a series of technological stand-ins for the sake of survival (Greek myth identifies them as the gift of the Titan Prometheus). In Gehlen's picture, the fact that technology is essentially utilitarian and belongs in the realm of necessity rather than freedom comes out even more clearly than in Marx: simply put, the human race would be incapable of surviving without its manifold technological creations. However in Gehlen's work it is not just the technical means of production or self-protection that come within the compass of this view. The social institutions which Protagoras still thought of as a form of political technology gifted by the gods are subject to the same analysis; for Gehlen, social norms and forms, become thoroughly utilitarian as well.

Technology for use, technology for pleasure

As we have said, the tendency to define technology in terms of use-value has its origin in the *theory* of technology rather than the actual history of technology. If we can point to the machinery involved in mining, metallurgy, building and warfare as early examples of technology for use, we should also recall that since antiquity there have been other more playful, less purpose-built forms of technology – machines that existed to edify and entertain. The difference between the two types of technology is already clearly articulated in a technical manual of the early modern period, Salomon de Caus' *Of Strong Motions: Description of Divers Machines Both Useful and Amusing* (*Von gewaltsamen Bewegungen: Beschreibung etliche so wol nützlichen alß lustigen Machiner*[5]). The book appeared in French and German in 1615 during de Caus' tenure as an engineer and landscape designer at the Heidelberg court. His most famous creation belongs to the same period; this is the garden supported by walls to one side of Heidelberg Castle that rises to a level with the surrounding hillside (*hortus palatinus*).

The word 'machine' in the title of de Caus' book makes it clear that he is talking about technology in a classical Greek sense – technology as *mechanike techne*, the collective term for devices designed to outwit nature into performing feats of movement that she would hardly perform on her own. The aspect of technology that de Caus is interested in is *para physis*, beyond physical nature. His discussion thus centres on violent movements – motions that man exacts from nature by force. De Caus thinks of technology as an art of bringing nature to do what is contrary to her basic ways, and he divides that art into two basic types: the useful and the pleasurable.

The frontispiece of *Strong Motions* ascribes each of the two types to a technological mastermind – Archimedes becomes the metaphorical father of technology for use and Hero of Alexandria the father of the technology of pleasure. This certainly involves de Caus in a stylization, for the historical Hero was far from impractical; although best known for his automata, his work for the Greek stage and for an ornamental fountain bearing his name, which utilizes heat to shoot water high into the air, he also built a crane of sizeable proportions, wrote a treatise on the art of fortification and developed theodolites for use in the construction of canals and tunnels.[6] Archimedes, on the other hand, is known exclusively for technical creations of a utilitarian kind. His law of leverage put the design and use of scales on rational foundations for the first time in history, while his discovery of the principle of displacement made it possible to determine specific weights for the first time. Even more practically, he is supposed to have set enemy ships ablaze during the defence of Syracuse using gigantic burning glasses.

As the notional founder of useful technology, Archimedes is depicted on de Caus' frontispiece with a set of scales and a basin of water with a crown floating in it; in the background, we see the defence of Syracuse. Hero's attributes, on the other hand, are a suction pump, flutes and the eponymous fountain. Both figures are allocated a patron from among the Greek gods, who stands in an overhead niche diagonally opposite his earthly representative: for Hero it is Hermes/Mercury, the god of communication, with his staff and pan pipe; for Archimedes it is Hephaistos/Vulcan, the god of smiths and metallurgy, equipped with a bellows.

Throughout *Strong Motions*, the distinction between technology for use and the technology of pleasure remains strict. De Caus deals, for example, with pumps and their use in mining; as in the fifteenth century work of Agricola,[7] they appear in a separate chapter on mining instruments. The technology of pleasure takes pride of descriptive place however: under the heading of *lustige Machiner*, de Caus deals, for instance, with automata that give life and interest to gardens – the moving figures of gods, birds that stir into song at dawn, water-powered organs and the like. Infact, there is a solid reason for the emphasis on technological ornament and technological entertainment for the wealthy; in de Caus' day, the technology of pleasure was a largely open, creative field, whereas more utilitarian forms of art and craft were largely in the hands of trade associations. Age-old tradition and, in part, explicit guild regulation governed everything a craftsman did, including the materials that he used and how the finished product was to look. For engineers of the seventeenth century – and the word engineer, we recall, means literally an inventor, an *inventores* – there was little or no room to apply their skill or imagination in fashioning devices for everyday use. Instead, it was at court that machine builders could take on the true mantle of *inventores* for the courts of the day could simply disregard guild regulations. In his preface to *Strong Motions*, de Caus praises Charles V of France and Henry VIII of England for 'having again stirred the free arts to new life'.[8] As well as military technology, it was the technology of pleasure that occupied the 'free artists' of the courts, like de Caus.

In picking out a particular non-utilitarian strain of technological development, we have uncovered a particular source of modern technological history. While mainstream histories of technology take their lead from useful forms of technology and thus tend to assume that the emerging bourgeoisie was the main social carrier, and industry the main economic framework, of historical developments, in looking to the princely courts of the seventeenth century, with their economies of luxury, we begin to see a second source of change. The court was the origin of a large amount of technology whose point was sheer entertainment. Yet the inventions that were pioneered in and around the courts beyond the reach of the guilds cast a light well beyond the realm of entertainment and pleasure. Because of the absence of guild restrictions, the new technologies of pleasure were also sometimes more technically innovative than the functional devices developed under more traditional conditions.[9]

As a rule, the technology of the early modern European court tended in the direction of precision engineering and deliberate mechanical complication. The mannerist aesthetic that marked the courtly crafts of the period was distinctly noticeable in the technical domain; it seems also to have inspired the remarkable fetish for automata and small-scale technological mimicry that shaped European tastes down to the eighteenth century,[10] culminating in Vaucanson's mechanical duck – a creature that not only walked and quacked but also ate and, to all appearances, digested what it ate as well. However, the greatest of the period's large-scale technical works were of courtly origin too. In its day, the Great Machine of Marly, near Versailles, was the biggest water-powered power plant in the world. Driven by 14 eleven-meter high water wheels, its purpose was to pump the waters of the Seine to an elevation of 163 m in order to supply Louis XIV's chateau at Marly and the fountains of Versailles. This was a technical feat without any economic use whatsoever. Without the economy of luxury in evidence at the court of the Sun King, it would have been inconceivable.

The technology of the court was focused above all on *curiosities*; its aim was to surprise, to turn up the improbable – to bring to light precisely what the everyday experience of nature would least expect. We should think of this side of technological development in connection with the art collections, the *Wunderkammer* and the court bestiaries which also made their appearance around this time, and which were designed to display anything that was disconcerting or exotic rather than nature as such.[11] Deformities and monstrosities were worth more in such collections than the most magnificent specimens of a well-known species. And yet their significance extends still further. We need to recall that it was in this courtly milieu that modern natural science actually came into being. Sixteenth-century science, in short, was not interested in nature in its customary naturalness, but precisely in what was *unexpected* about nature. Its leading minds wanted to investigate nature under conditions that teased out phenomena that looked, on the classical picture, as if they ran contrary to nature; the new science was looking to uncover what was *para physis*, i.e. what appeared only under technical conditions. This, in fact, is the very essence of Galileo's revolutionary new approach to scientific discovery. (Newton, in a similar vein, had famously had to defamiliarize himself with natural appearances – apples falling from trees for example – in order to arrive at explanations in well-formed physical terms.[12])

However the spirit of *curiositas* didn't centre wholly on the court, even if wealth and the absence of guild tradition meant it could be indulged most lavishly there. A broader impulse to representation was at work. The art collections, the *Wunderkammer*, the bestiaries, the waterworks and theatrical marvels, the automata that filled gardens and grottos, were part of the brilliant spectacle of the court. But it was the *relationship* between scientific curiosity and the courtly impulse[13] to display that gave shape to the early phase of modern science rather than curiosity or courtly pomp alone. From demonstrations of new telescopes and microscopes, to the anatomical theatres of the early universities through to the public staging of experiments, modern science bears the mark of a certain theatricality. Informing the public went hand in hand with entertaining it, and the new science, which set so much store on novelty and curiosity, had its place at annual fairs as much as it did in the secluded chambers of princely courts – as we see very clearly in the burgeoning seventeenth-century field of pneumatics and in early research into electricity and the vacuum. Probably the most impressive of the early experiments with vacuums was conducted by Otto von Guericke in 1656 or 1657; it set out to show that 16 horses were unable to prise apart two metallic hemispheres separated by a vacuum (or something we would say today approximates to a vacuum). The fact that Guericke presented his experiments directly in front of the parliament building at Regensburg (where he was posted for a time as a diplomatic envoy of the city of Magdeburg) is a good indication that here, too, representation in several senses of the word was part of scientific activity.[14] To be sure, scientific issues were at the heart of Guericke's project – air pressure, the fluid-like properties of air, the existence of the vacuum, and the like. Yet what Guericke was demonstrating was first and foremost an exciting new technological form; the device that created his vacuum was something that, at least at the time, was without any use whatsoever.

Though Guericke's experiment itself took place in the mid- to late-1650s, the results were only published in 1664 in a wide-ranging work, *Technica Curiosa*, by the senior Jesuit Caspar Schott.[15] Both the title of the book and the bulk of its contents reveal that Guericke's research was indeed scientifically serious,[16] but that his attention as an experimentalist centred on the astonishing facts that human ingenuity could produce by technical means. Curiously, in underlining that very point in his preface to the reader, Schott simultaneously reproduces the division between technology for use and the technology of pleasure already known to us from the work of de Caus:

> Nothing better distinguishes the excellence, the inventiveness and the diligent spirit of a human being than the astonishing technical achievements of art brought forth by the individual for the use or amusement of his species.[17]

Today, Guericke's experiments with air pumps are regarded as much more than a form of technological showmanship and they come down to us in an account of the formation of an expert circle devoted to experimental philosophy within the London Royal Society. However, in Schott's text, they make their appearance alongside designs for a perpetual motion device, a proposal for a universal language and ingenious plans for fountains designed to raise water above the level of the sources that feed them. None of Guericke's schemes is designed for use, but for the amazement and amusement of the general public. And so it goes in general. Vaucanson's mechanical duck, the waterworks of Marly and Guericke's metallic hemispheres all belong, partly or wholly, in an historical category of their own: that of the technology of pleasure.

Aesthetic economies

The fact that historians of technology who see technology from the standpoint of use point time and again to the exorbitant and – in their opinion – wasteful costs of the technologies of pleasure is symptomatic of a wider problem of intellectual interpretation. Commentators already level the charge at ancient technologists such as Hero. In relation to Hero's automata, Wolfgang König notes drily that:

> [Hero's] automata need to be seen in the context of the Alexandrian court of the day; the task of their creator was to contribute to the entertainment of courtly society by creating surprising effects. The Alexandrian machine-builders were given the chance to build and test complicated devices free from economic pressures.[18]

Troitzsch has this to say about the Marly waterworks:

> This astonishing edifice was on an enormous scale, not just by the standards of the day, and with a price tag of almost four million *livres*, construction costs too were a daring extravagance ... Questions of money were only of secondary importance; courtly employers were prepared to pay enormous sums to satisfy their wishes.[19]

In a similar vein, Fritz Krafft is unable to resist pointing out that Otto von Guericke's expenses for his Magdeburg experiment – a sum of 20,000 thalers – were grossly disproportionate to the money the project stood to bring in.[20]

Does the technology of pleasure need to be rethought from an alternative point of view? The one-sided reading of courtly technology from a budgetary perspective would seem to indicate as much.

There are certainly forms of economic life that run contrary to our normal notion of economic utility: the economics of gift giving and the American Indian institution of potlatch are two examples. More importantly for our purposes, for more than 100 years there has been an alternative theoretical view of the economics of capitalism – an alternative, that is, to mainstream economic analysis from Adam Smith and Karl Marx to Max Weber. From the mainstream standpoint, there are two crucial points here. The first is that capitalism is a creation of the bourgeoisie, the second that capitalism presupposes an economy of scarcity – a consumption-based economy in which consumption is subject to material or moral constraints. Both the accumulation of capital necessary in a capitalist economy and the way the use of capital stimulates further production result in this view from the fact that profits are not simply consumed but are also reinvested in the capital-formation process. Famously, according to Weber, the cultural backdrop to the individual behaviour that is necessary for such a system is provided by the Protestant or, more specifically, Puritan work ethic.

For a *counter-concept* of capitalism, we might turn to authors like Thorsten Veblen, Werner Sombart and Georges Bataille. In his *Theory of the Leisure Class* (1899), Veblen charted the dissolution of puritan ethics; Veblen's argument was implicitly that because the leading strata of the capitalist economy were increasingly given to *conspicuous consumption*, the strictures of the economic mainstream failed to apply.[21] Sombart was to go one better in his *Love, Luxury and Capitalism* of 1913, tracing capitalism's very origins back to courtly luxury and wasteful consumption.[22] Bataille, we might say, produced the *pièce de résistance*, bringing the development of heretical theories of capitalism to a close. Generally inclined as he was towards a post-Victorian ethic that gave positive emphasis to waste and excess, in his economic thinking Bataille came to the conclusion that capitalism is marked by alternating periods of accumulation and wasteful indulgence, rather than by any straightforward trajectory of accumulation and growth.[23]

Now it should be said immediately that none of the theories that we have just labelled heretical entirely contradict Marx or the mainstream. Consumption obviously represents an important aspect of the capitalist economy on the Marxist reading, as it does in later Keynesians and neoclassicists. Sombart is, nonetheless, quite right to note that the demand generated by the courtly taste for luxury was one of the sources of early capitalist manufacturing and later of fully industrialized production, and that the process as a whole was at odds with capital accumulation among consumers. On the other side of the debate, Marxist writers have hardly been unaware that the *overaccumulation* of capital and the destruction of capital (e.g. in war) were part and parcel of capitalism too. However, the most important result of the whole debate between orthodox and alternative theorists surely has to do with the significance of *excessive* consumption for the functioning of capitalist economies generally – a significance which could only come into view once historical developments had entered a phase in which basic needs were being satisfied, and indeed satisfied easily enough for the economic growth so necessary for capitalism to be called into question.[24] Under such circumstances, economies necessarily began placing a premium on needs that were actually *intensified* rather than satisfied in being met. With that in mind, we can now make a terminological distinction between the two types of need by calling needs of the self-amplifying kind desires (in a limited sense of the word). And we could term the form of capitalism which places a premium on such desires an *aesthetic economy*.[25]

The justification for using a special term here lies in the fact that in the later of the two phases of capitalist development a new type of economic value emerges – something that is, in a certain sense, a hybrid of exchange-value and use-value: we will call it the *spectacle-value* of a product. Products, in short, are by this stage no longer simply goods to be used or consumed, they become status symbols or exhibition pieces of sorts.[26] Consumption itself changes its meaning as the consumption of (what were formerly) luxuries becomes routine and widespread – a process which takes place *within* the framework of a capitalist society, not by pushing beyond it, as Sombart had argued.

Yet here we need to take a step beyond Veblen, Sombart and Bataille, for what is striking is that all three fail to note the particular role technology plays in a consumption-driven economy. Sombart, for example, talks at length about velvet and brocade, silverware and giftware. Louis XIV's elaborate building programme comes briefly into view. However, the Sun King's expenses on automata, ornamental waterworks and, in particular, the Great Machine of Marly go unnoticed, as do the French court's vast expenditure on fireworks, i.e. on *pyro-technics*. None of the three unorthodox theorists of capitalism pay much attention to the existence of distinctly unproductive forms of technology: technology that in economic terms represented sheer waste, served no other purpose than deliberate excess, or was consumed for the sake of pleasure in technology itself. To get an idea of the importance of the field, we need to recognize that many forms of technology began life with no other function than display, prestige, sport or public amusement. As in the case of air travel, electronics and the automobile in the twentieth century, the early technologies of pleasure had been in existence for a considerable length of time before any sort of useful or economically profitable application for them came into view.

It could well be that the pattern of these historical developments is only becoming clear to us today because the rich industrialized countries of the West now find themselves caught up fully in the dynamics of an aesthetic economy; in the developed world, contemporary capitalism seems to suffer from near-permanent sluggish economic growth, which we try to make good by boosting consumption by all available means. In contemporary capitalism, the consumption of high-tech gadgetry has become a central feature of the economic system. What that means in our terms is that our second major type of technology, the technology of pleasure – which historically has tended to play an important but still marginal role – has taken its place at the very centre of economic life.

Technology and consumerism

In the midst of the essentially aesthetic economy that has developed in the Western world, the difference between technology for use and the technology of pleasure seems to be vanishing. As we have seen, a new type of economic value has come to define the aesthetic economy: the spectacle-value of products. The latter, however, is at bottom a composite of use-value and exchange-value; it makes the features of the product that serve the purposes of marketing into a quasi-substantial aspect of their use-value.

We can see a related composite of utility and pleasure in new high-tech products; they become useful precisely by becoming objects of pleasure and fun. Of course, the economic utility of the products in question accrues mainly to producers and marketers; thus the high-tech products of the entertainment industry have come to make up a major sector of the economy. The lavish arrays of equipment necessary for the consumption of audio-visual material are a case in point. We could think of them as a kind of throwback to the vast courtly apparatus of amusement that European monarchs surrounded themselves with, bearing in mind, of course, that consumption has been almost wholly democratized in the period between 1650 and today. Yet the same hybrid of utility and pleasure is not just evident in isolated corners of society, but in much of the basic technological furniture of the contemporary world, including computers and mobile phones; although both were conceived originally for use in one or another narrow economic or scientific setting, the fact that they also have non-functional applications, above all for purposes of play, has made it possible to develop them into objects of mass consumption. Technological mass products like Game Boys, X-boxes and electronic gaming machines, which were designed explicitly as high-tech toys, have been developed alongside these crucial crossover products.

In recent times, games, television and the consumption of DVDs and music have become something like basic needs of the broader population as our labour-oriented society of old has been transformed into a leisure society – to speak with Veblen. In the 1970s, Herbert Marcuse had already shown that our new – nowadays not so new – leisure societies were far from the ideal of freedom that Marx thought would be realized once labour was no longer necessary to satisfy basic human needs.[27] The capitalist economy ensures, however, that leisure is governed by the same principle of *productive achievement* as the world of work. In concrete terms, what that means is that human beings' needs – the need to communicate with others, to listen to music and to take part in the events of the world around them (e.g. vicariously by means of images) – are transformed into insatiable desires. The transformation takes place moreover in and through an upwards spiral of technological improvement. The multifunctionality of high-tech appliances and the interconnections between them, are constantly upgraded, as are picture and sound quality, so that the television or stereo or mobile phone we already have becomes unusable, even if it can still, in theory, do what it was designed to do. Everything suffers from what Marx called *built-in obsolescence*: the television, stereo or mobile phone becomes antiquated because it soon fails to live up to current performance standards, ceases to be compatible with other new gadgets or can no longer process data from the latest generation of data-streaming technology. Society comes to be caught up in full-blown technological consumerism in the process. As we have suggested, technology today has become much more than a means of production or entertainment; the emergence of the entertainment industry within a wider aesthetic economy has led to a situation in which technology is consumed for its own sake.

The latest generation of mechanized gadgetry gives us a picture of technological consumerism in its purest form. What we are talking about is nothing less than a third great historical wave of enthusiasm for automata that parallels and, in many ways, trumps the obsessions of the Alexandrian and Baroque periods. Following the post-war popularity of all kinds of motorized and remote-controlled toys, the products of so-called *artificial-life technology* now dominate the market. A typical success story was the *Tamagotchi* – an egg-shaped mini-computer with a display simulating the life story of a living creature which rapidly became a worldwide sales phenomenon and sparked more than a touch of commodity fetishism. *Tamagotchi* users were charged with the task of looking after these virtual beings; they were responsible for the development and health of their computer-generated pets, which required constant care over the course of weeks. Part of the aim seemed to be to inculcate parenting and teaching skills of sorts: the individual *Tamagotchi* developed into more or less likeable characters and lived longer or shorter lives, depending on the quality of the care that they received. The fact that the lives of the *Tamagotchi* were finite seems to partly explain their overwhelming popularity among a generation of consumers. The *Tamagotchi* were programmed to make resetting impossible – a design feature that contributed, whether intentionally or not is difficult to say, to the respect for electronic life that came to be demonstrated by so many of the owners of these curious toys.[28]

The *Tamagotchi*, too, eventually fell prey to built-in obsolescence. Product lines which make use of more sophisticated robotic technology and which aim to speak to the pleasure many people take in so-called *artificial life* have now been with us for quite a while. One gadget which perhaps deserves special mention is an electronic dog that went by the commercial name of *AIBO* (Artificial Intelligence roBOt). Like the *Tamagotchi*, the new generation of motorized canines required care from their owners. Where they differed was in being endowed with a definite intelligence which developed in response to human interaction and made possible a distinctly personal sense of interconnection. *AIBO*'s creators, we note, made sure that the technology of pleasure in this case had some classic practical functions to boot – one's canine friend functioned, somewhat bizarrely, as a calendar and an alarm clock as well.

It is curious that in capitalism's contemporary phase the marketing for technical gadgetry, where possible, tends to emphasize use-value even where this is almost wholly notional. Our economies, it would seem, have made the switch to their present wasteful, consumption-driven phase rather half heartedly. Just as we cling to the fiction that our society is defined by work, so too we hold fast to the notion that technology can be understood in terms of use. Technological novelties like smart clothing and smart houses are praised as practical, though they clearly speak to little more than our pleasure in technology for its own sake. Clothes which independently monitor the heart rate or the skin temperature of their wearers are surely only of real practical use in conditions of war. Likewise, a house which lets its owner run a bath while sitting in his car in the driveway is surely equipped to the point of technological excess. But having the clothes or the house is, quite simply, a lot of fun. The high-tech devices and systems, like the aesthetic gadgetry in the wider aesthetic economy, become exhibition pieces, props for creating casual effects on the stage of theatrical self-display. Rather than satisfying basic needs, they exist to meet desires that consumption ultimately doesn't satisfy but magnifies.

Conclusion

It would be possible at this point to object that the distinction between technology for use and the technology of pleasure leaves one major form of technology out of the picture: military technology. Military technology is indeed one of the central dimensions of technological innovation – a place it occupies because of two things it shares with the technology of pleasure: it was never subject to guild restrictions nor is it expected to bring in profitable returns. On the other hand, perverse as it may seem, military hardware has often been counted among the useful forms of technology. It thus seems tempting to put it in a category of its own, somewhere in between useful and amusing technology. The destructive capacity of the technology in question should make us wary of any such interpretative move, yet the suggestion alone throws light on one aspect of military technology that has been essential throughout history: like the technology of pleasure, the latest products of weapons makers have come into their own for purposes of representation; nations that have been active in their development have gained notably in prestige, even when their advanced weaponry has never been used to fight actual wars. Demonstrations of technical military capacity – everything from parades to airshows to the actual detonation of atomic bombs – have taken on some of the qualities of national festivals. If military technology seems, at first glance, to belong purely within the economy of consumption, public demonstrations of military might can look disturbingly like another of form of technological entertainment.

Yet as we have argued, in its present advanced phase, capitalism is fast blurring the boundary between the different types of technology. (Maybe we should say that, under contemporary economic conditions, nearly all new forms of technology come to have some of the traits of the technology of pleasure as they cease to serve the satisfaction of basic human needs.) In any case, in important fields such as communication and aeronautics, contemporary technological developments can no longer be even remotely understood from the standpoint of utility. The number of forms of technology that belong in the grey zone between technology for use and the technology of pleasure seems set to grow.

Technology in the life of an everyday philologist

Introduction

Victor Klemperer's diaries[29] are no *journal intim*; while their author's unique sensibility – his very manner of being affected by things – clearly determines what he writes about, he rarely gives himself up to reflection in the full sense of the word. Klemperer, for example, makes critical note, time and again, of the coldness of his reactions to the misfortunes of others; he asks himself if his mildly envious lack of enthusiasm for his wife's choice of career is to blame for her feelings of depression. Yet his own inner being, his own personal development, is not his main theme. Walter Nowojski's choice of title for the diaries dating from the period 1933 to 1945 – *I Shall Bear Witness* (the phrase is Klemperer's own) – is surely a wise one, for Klemperer became increasingly conscious of the documentary significance of what he was writing as the distress of Jewish life in Nazi Germany took an ever grimmer turn. However the title of the volume covering 1918–32 is better suited to the diaries as a whole; Klemperer's watchword, which his editor takes up here, *Leben sammeln, nicht fragen wozu und warum* ('Collecting life's fragments without asking why'), sums up the project of this kind of writing perfectly. Klemperer is the chronicler of his own life. What is worth writing about is no more and no less than what each day of life brings.

Klemperer's general sense of himself, and of life, is rather cautious. He certainly has no vivid sense of life's possibilities, let alone a goal he is passionately driven to realize. In the end, he becomes a specialist in Romance language and literature, yet at no stage of his decade and a half at Dresden University does he begin to identify himself with his chosen academic field. Certainly, the act of writing – both his studies in literary history and his efforts as a diarist – define a biographical trajectory that Klemperer perseveres with, that he indeed remains true to with admirable self-discipline in situations of direst difficulty. Yet life remains something he experiences passively. (Again, this is not to say he is anything but an extraordinarily keen observer: every page of the diaries registers something of the flux of public life, the mood of his contemporaries – everything down to their speech habits – with an immense sensitivity.)

What are Klemperer's main themes in the diaries? What exactly does he find noteworthy? First and foremost the confused web of human relationships: relationships with friends, relatives, neighbours, work colleagues, politicians, artists, writers, servants and tradesmen. The number of entries that appear in the register of names that Nowojski has appended to the diaries covering the years 1918–32 brings out the sheer breadth of Klemperer's social life – something we might find it difficult to imagine today: 2,500 separate figures find their way into this volume alone. As well as accounts of relationships with other people, we find details of the Klemperers' travels, problems that Victor has at work, thoughts about his literary work, and about his cultural experiences in the theatre, the cinema and at concerts. Politics comes to preoccupy Klemperer when it intrudes into everyday life and in the end completely permeates the world around him, almost obliterating his immediate lifeworld. In the years after 1933, one sees it most strongly, and for obvious reasons, yet something similar had already happened once before, amid the confusion of 1918 and the calamity of the German defeat in war. As good an indication of the range of Klemperer's interests and themes as that may be though, we can add one more item to the list: technology. Like politics, Klemperer experiences technology as something that intrudes into everyday life and, from time to time, completely defines or redefines it.

Until now, the importance of technology for the Klemperer of the diaries has been overlooked, understandably so given that it was their everyday sense of political life that originally brought the diaries to broad public attention. In any case, one hardly expects Klemperer to have much to say about technology. What would pronouncements about technological change be doing among the musings of a melancholy philologist – someone incapable of manual work, and without technical or scientific training of any kind? Given his background, Klemperer's open-minded attitude to new technology is all the more striking; at times, what seems to come over him is a sort of passionate involvement in *modernization itself*. It is precisely the fact that he comes to technology as an uncomprehending outsider that makes his testimony all the more valuable; what the diaries conjure up is the human experience of technology, a vivid picture of the successive phases of the technification of everyday life. Here, too, Klemperer proves to be an immensely sensitive and acute chronicler. His comments on the flight of the zeppelin on 5 October 1930 are as good a sample as any:

> I feel nothing of the enthusiasm of years past. The opposite in fact, a feeling of embarrassment. The way the airship laboured into the wind, never seeming to shift from the spot! What struck me most was when it was directly above us, not far off the ground: what a monstrous object, what a strange mass, what a waste of machine power just to get the tiny (relatively tiny) cabin into the air! Like building an ocean liner to haul a single suitcase – an ocean liner you could only use to haul a single suitcase. I was immediately convinced that the future belongs to the aeroplane. The zeppelin is fated to be a museum piece.

We're electrifying the lot – kitchen included

What social historians know about the technification of household life in the interwar years is usually based on what they can glean from official statistics, advertising, company archives and the agendas of contemporary housewives' associations. What people's first-hand experience of new technology was like remains hidden from view. Where indeed would one turn if one wanted to find out about such things? They are normally not important enough in literary terms to find their way into novels, poetry or plays except as secondary details. Most literature represents them, if at all, in a reconstructive sense: scenes in novels, we might say, provide a snapshot of *the result* of technological development, whereas diaries document the piecemeal stages of historical change, change as a sort of open-ended adventure.

Victor and Eva Klemperer's experiences with their new electric stove are typical here. After a wave of initial enthusiasm, their modernizing passion leads rapidly to disillusionment.

> Eva is in raptures. We're "electrifying" the lot, kitchen included. I am buying the larger appliances in instalments. It means paying more, but this way the financial burden is spread bearably over time. (12 September 1934)

After decades of living hand to mouth and several years of arduous planning, Klemperer and his wife have finally borrowed money to build themselves a house. Klemperer himself is aware of the ambivalence of this seemingly solid achievement: he knows the chance to own their own home is opening up to them because of the economic policies of National Socialism. (Germans had recently been forced to liquidate their foreign assets and convert the proceeds into German *Reichsmark*.) With a surplus of hard currency on her hands, the sister of one of Klemperer's colleagues is more than happy to invest her money in the Klemperers' plans for a new house, as we read in Victor's entry for 14 July 1934. The house goes ahead, but still manages to exceed the couple's means. Victor presses ahead and has electricity connected throughout the house, including the kitchen. His attitude is that he's starting a new life, so he wants it to be modern – even if the modernity comes in instalments, like the house itself. Disappointment follows hard on the heels of excitement. A week after moving in, Klemperer writes:

> The biggest of the dubious novelties in the new place are the electric stove and the central heating. The oven is almost totally unusable – one minute it completely overheats, the next minute it gives out. The cooking plates take twice, maybe three times as long to heat up as a gas stove. (14 October 1934)

Three months later, we read:

> Today an exorbitant electricity bill. The electric units [in the new house] are a complete disappointment. Cooking proceeds at snail's pace and costs far too much. (16 January 1935)

Then barely six months later, they give up completely:

> A lesson in home economics. I'm relieved now we've had gas re-installed. The electric cooking units which got so much press (we fell for it) are a straightforward swindle: they're too expensive and too slow for use in a small kitchen. (30 May 1935)

The basic pattern of experience here – the mixture of expectation and enthusiasm with which they take to new technological offerings, the disappointment, the return to the old – can surely be explained in terms of something more than the waxing and waning of the Klemperers' taste for modernity: this particular new appliance obviously needed to be technically fine-tuned and integrated affordably with energy supply. But the episode is also something else, an index of the role technology was held to play in creating new forms of private and public life, a testament to a general willingness to take on board the new and a testament to the lack of preconceived ideas or latent suspicions about technology characteristic of the interwar years.

Can you believe – I'm dancing

Victor and Eva were undoubtedly sociable people: social interaction made up the greater part of their life, even in the prison-like *Judenhaus* in the depths of the Nazi years. Until well into the 1920s, the Klemperers epitomize a classic middle-class form of sociability. They hold dinner parties, they try out the games that people of their social set are playing, they gather with friends and do a little music making, recite poetry or read novels in the round, debate the politics of the day or talk shop. Then, on 5 November 1926, Klemperer announces something new: 'the big event: *the gramophone*'. Social nights at home are about to change dramatically, for while the couple have kept up some of the habits of their earlier bohemian days together – visits to Dresden clubs and bars went on after Victor had taken up his post at the university – now the spirit of the Roaring Twenties has found its way into their middle-class home: per the medium of the gramophone. Klemperer buys his first LPs the very same day, setting the scene for another radical break with middle-class tradition. What he finds himself listening to collides with his conventional sense of music and taste – 'mainly American stuff, mainly Negro', as he puts it. What follow are heady days for the couple and their friends:

> Since then it's been guests and dancing every night. Can you believe – *we're* dancing! Eva is even dancing *well*. I find it a bit embarrassing the way you have to get sexually so close, intertwine your legs – it's like a danced form of coitus. (5 November 1926)

One hardly expects the initial ecstasies to last, and they don't. But 12 days later, the impressive new gadget is still dominating their lives (together with Victor's hopes for a promotion):

> We're having more people over than usual, we dance until late into the night, I'm continually tired.

By May 1928, the gramophone's fascination seems to have well and truly worn off. Klemperer writes:

> Dancing, jazz – the gramophone has fallen almost completely silent at our place. (9 May 1928)

Then, 7 years later, as the influence of National Socialism is making itself felt throughout every aspect of daily life, the earlier musical experiment awakens memories – in this case a kind of longing for the cosmopolitan past represented by their joyous earlier encounter with the *Zeitgeist*:

> When the Wieghardts are over, Eva sometimes plays some of the old hits. These tangos, this nigger music[30] and the other international and exotic stuff from the days of the [Weimar] Republic all now have an historical value. They really move me, as well as making me feel genuinely bitter. They're full of the spirit of freedom, full of cosmopolitanism. In those days we were free, we were Europeans, we were human. Now though – (7 February 1935)

Just a point in space

On the whole, Klemperer and his wife only rarely make the switch to new forms of technology out of economic necessity or to rationalize their day-to-day living arrangements. What motivates them more often than not is curiosity – a desire to follow, and take part in, the life and times of the society around them. This is precisely why they get so emotionally involved in the process: their encounters with technology take on the quality of genuine life experiences. The two plane trips that the couple take in 1927 and 1930 bring the point out especially clearly. The first is a flight from Dresden to Leipzig. Klemperer attempts to mentally fix as many details on the page of his diary as he can – he tells us, for instance, that 'the plane's cabin is much smaller than the passenger compartment of a cab'. Although coloured by the conscious sensation of entering a new world, the experience remains primarily a bodily and aesthetic one. The small plane that the couple flies in accelerates sharply and is buffeted about in mid-air, leaving Klemperer feeling nauseous. But, in the end, he overcomes the physical reaction: the aesthetic aspect of the little adventure carries the day. 'It's grand, mildly distressing (the pressure in one's ears) this feeling of being up there in "space", the feeling of being hurled forward ever greater distances, an incomparable experience' (5 August 1927).

Klemperer has some fun at the expense of an older petit bourgeois couple who are simply there to try out something new, and who demonstrate 'an impressive thirst for knowledge, a certain heroism which is both comical and curious' – though the description fits him and his wife equally well. Like their fellow travellers, the Klemperers belong to the second generation to take to the air – they are yet to see flying as an ordinary means of transport or as a rational necessity of everyday economic life, as it was to become in the post-war years, though they no longer see it, as for example Saint-Exupéry had seen it not so many years before, as an exploit on the very margins of civilized life. For the Klemperers, flying is a matter of radical novelty, and that holds true for both their 1927 trip to Leipzig and the first trip that they take for more than curiosity's sake in 1930 from Dresden to Stettin (the point of departure for their traditional summer holiday on the Baltic). In 1930, Klemperer again stresses his bodily experience of space, though this time it is tinged by an awareness of danger. Clearly, accidents have been at the forefront of public attention in the time leading up to the trip:

> I kept saying to myself: In an aeroplane I'm *perceptibly* in the exact same position I'm *imperceptibly* in down below: I'm a single unsteady point in empty space, exposed to the cosmos' every breath. (9 August 1930)

In the midst of runaway inflation earlier in the decade, the couple had already thought of taking to the air. When connecting flights between Berlin, Danzig and Königsberg had begun in 1923, Victor had noted:

> Eva is completely taken with the idea. This morning she immediately started making enquiries. The trip doesn't cost much more than a second class ticket on the express train, roughly 400 000 marks. (14 July 1923)

But when they get to Berlin a week later, the price of the fare is already 2.4 million marks. They are forced to give up their original plan to fly, which is only realized years later, once the galloping inflation of the mid-1920s comes to an end with the introduction of the *Reichsmark*. The 1927 trip from Dresden to Leipzig ends up costing 20 marks.

Auto, Auto über alles

All things told, the Klemperers' experiences with car travel are a kind of tragicomedy that would be worth recounting for its own sake. What starts out as curiosity in Victor's case reaches the veritable heights of passion – only to fall foul of his complete impracticality (he is ham-fisted behind the wheel as well as being plagued by peculiar bad luck). The actual story lasts just three years, from December 1935, when he learns to drive, until December 1838, when German Jews are banned from the roads. But his lively interest in driving is already detectable much earlier in the diaries in his reports of car trips with wealthy friends. Of the journey by night from Görlitz to Dresden in the company of Hirche (an industrialist), Klemperer writes: 'The trip set my nerves jangling – but what an experience'. First and foremost, it is an experience of *speed* – a strange mix of conscious awareness (of how fast he is moving) and a trance-like sensation of bodily ecstasy. As they race through the night, Klemperer keeps his eye fixed on the odometer, yet still seems to give himself up to the sheer physical dynamics of the experience:

> I couldn't always see clearly what was around us, everything swam before my eyes, everything danced. The driver was completely calm and assured – I was in a state of dazed intoxication. Keyed up, then nodding off from time to time, often completely bewildered by the way we coursed through the night yet still enjoying every minute. On and on we flew.

A few years before, in May 1923, Klemperer had been offered by a ride the wealthy magnate Thiele, and he had jumped at the opportunity. Apart from the evocation of a bodily sensation of speed, probably what will strike the contemporary reader most about the scene as Klemperer paints it are the ceremonial niceties that the trip involves – 'They helped us into leather coats, put caps on our heads, adjusted our driving goggles':

> But the landscape around us didn't appeal to me as much as the special clothing and the trip itself in such a fast-moving, wonderfully elegant open-top. Ah, the speed – mostly around 40km/hr, a couple of times up around 60km/hr. I leaned forward and read the speedometer over the shoulder of the (entirely respectable and workmanlike) chauffeur. When the car reached speeds of about 50km/hr an almost surreal sensation of speed comes over you, a tension in all your limbs. (14 May 1923)

Klemperer's last comment is probably the most difficult to understand today, not just because the speeds that exult him seem – almost – pedestrian by today's standards. Einsteinian relativity tells us that speed as such is completely imperceptible – and yet there is no denying that Klemperer's are real experiences. (Nineteenth-century reports from the early decades of train travel make much of the sensations of nausea that regularly accompanied that new experience of speed too.) To judge by the diaries, it seems possible that Klemperer's sensation of speed was created by the sound of the car's engine, or that his sense of bodily orientation was thrown out of kilter by the flux of his external environment. For early generations of car travellers, it seems to be the case that the *conscious awareness* of travelling at speed was already a gripping physical sensation in itself.

The drama of Klemperer's love affair with the new form of travel actually starts at a time when buying a car of his own is something Victor knows to be an act of near madness. With the enactment of the Nazi Law for the Restoration of the Public Service, Klemperer is dismissed from his job on 30 April 1935. Shortly after, he loses his rights as a German citizen, along with all German Jews. The couple are already deeply in debt; the long-term task of paying off the house is just beginning. Now Klemperer's income is reduced by half. As a Jewish writer, his chances of being published are now negligible. Nonetheless, after realizing in his review of the year 1934 that he would 'now truly love to learn to drive', he begins with lessons in 1935. From the start, he seems to have little idea what he is getting himself into; learning to drive turns out to be nigh-on impossible (by this stage, Klemperer is in his fifties). The lessons proceed. Klemperer gets everything wrong, confuses left and right, hits the accelerator too hard and too often, and soon ends up feeling helplessly anxious. On 2 December 1935, he makes note in the diaries of a period of depression that has now lasted for weeks: its second most likely cause, after the trouble he is having writing the chapter on Diderot for his latest book, is the pain of learning to drive, 'the torment of driving, pointless in every way, completely contrary to nature'.

Does he mean his new hobby is contrary to his own nature or contrary to human nature per se? Straight after failing his first driving test, he enrols in a second course of lessons. This time, in January 1936, he passes the test. But his troubles are not over yet. Before he can buy a car, a garage has to be built – this is plainly the way things are done in the 1930s. But the garage by itself is already beyond his means. Official regulations now rule out a garage with a flat roof – the simple cheap variety – because this would be ungermanic (he had already run into the same problem when building the house). In March 1936, Klemperer announces the next step:

> Bought the car on 2 May. 850 marks, plus 18 marks/month tax. 6 cylinder Opel, 32 HP, four years old, open-top. (6 March 1936)

although K's entry for the same day already notes that 'the whole business with the car is an experiment in desperation' (clearly a comment about the couple's financial position). In a calmer frame of mind, he tells himself that the car is a justified expense because of Eva's ongoing health problems (the trouble she has walking is setting them back 100 marks a month in taxi fares). Yet obviously Eva's health is only part of his overall calculation. What tips the scales are the thirst for travelling that he has had since he was young – his lifelong need to get out of doors or out of town – and, even more importantly, his recent pressing need to escape the ever-narrowing circle of daily life in a world that by now has barred him from working and in which Jews are being slowly but systematically excluded from any sort of participation in public life. On New Year's Eve 1938, as he and Eva are forced out of their home and into the Dresden *Judenhaus*, he makes the point quite clearly:

> Until December we still had the car, we could still get about.

After listing the trips he and Eva have made, Klemperer reflects on

> all the little excursions, the freedom the car gave us to look after ourselves. Not to mention going to the movies, eating out. It was a little slice of freedom, of life, though maybe it was pitiably small – maybe we were right to think of it as slavery even at the time.

It is worth noting here that Klemperer already recognizes when he buys the car that it stands at the centre of a network of things and social transactions; contact with car dealers and mechanics, paying taxes and organizing insurance – all make him acutely aware of the surrounding world that the car is part of:

> What a curious kind of economic object a car is – the endless goings-on around it – it's a whole world in itself . . . (6 March 1936)

Familiarizing himself with this world turns out to be endlessly difficult. For months, the car seems to cause Klemperer palpable suffering, not just because of how much it is costing him, but because there is always something that isn't working; all in all, the new purchase presents him with a string of contingencies that he is almost completely unequal to. More than once, he runs out of fuel on the road, within months the speedometer gives up the ghost, the car refuses to start, the accelerator sticks and he has to drive with the brake on as well. At one point, the car shoots forward by itself and lands Klemperer on the pavement – on and on it goes – he runs into all sorts of things, he can't manoeuvre his way through the garage door, he wrecks Eva's flower garden, clips another car.

All these mishaps are interesting for a reason. Klemperer's tales of learning to drive make it clear enough that he has no real idea when and how to use the accelerator. Yet what becomes clear from his 1001 minor accidents is that Klemperer actually has no feel for handling the car *in any sense*. He seems unable, as it were, to *live his way into the life of a driver* and he is certainly totally incapable of treating the car as an extension of his own body, as an experienced driver does. However, while it would be easy enough to put most of his problems down to age, it is quite possible that everyone who had lived most of his life in a world without cars would have started off with similar problems – and in Klemperer's day that would still have been the vast majority.

To begin with, the car almost drives him to despair:

The car is eating away at me – at my heart, my nerves, my bank account. It's not so much a matter of my miserable driving or the occasional stir it creates, not even the trouble I have getting the car in and out of our place. It's the fact that the car is never working properly. There's always something out of order. I've lost all trust in the car. (12 April 1936)

Twelve days later, he writes:

The car is still taking over my whole life . . . I keep on denting the fender, I keep on knocking into the gate or the garden wall.

For the time being he tells himself to take heart: 'I have to persevere. Maybe I'll start enjoying this at some stage soon' (24 April 1936). By 28 April 1936, things are looking up a little: 'Gradually this driving business is becoming a little more satisfying. Maybe there's not that much to it. In itself it's a pleasure and a distraction'.

Once the car is in close to full technical working order, a real passion for driving seems to come over both Klemperers. They visit relatives, drop in on exhibitions, go sightseeing, take a trip to their beloved Kipsdorf (a health resort), then head on to Berlin, Frankfurt an der Oder and Leipzig. At the close of 1936, Klemperer calculates that he has driven 6,000 km in the course of the year. On May 21 of the same year, he had already been struggling for words to express what had come over him:

Auto, auto über alles, it's got to us completely, it's a veritable . . . *passion dévourante*.

On 5 July 1936, he tells us that the price of petrol makes longer trips prohibitive. His new passion isn't just devouring his bank account though, it is also eating away the energy he has for his writing:

The car is eating away at me. The work on Rousseau is just a stop-gap.

The comical tag *Auto, Auto über alles* is of course a reference to the – at that stage more and more aggressively charged – German national anthem, and shows that Klemperer is well aware of the politics of his new hobby. The fact that the new world

he is immersing himself in is also part of Hitler's Germany becomes unmistakable
when he and Eva find themselves caught up by chance in the opening of a new
stretch of the *Reichsautobahn*:

> Last Sunday too only a very short drive for reasons of economy, but peculiarly
> interesting. Half by chance we found ourselves on the new *Reichsautobahn* from
> Wilsdruff to Dresden, less than an hour after it was opened. There were still flags and
> flowers from the ceremony in the morning, a mass of cars moved slowly forwards
> at a sightseeing pace, only occasionally did anyone attempt a greater speed. This
> straight road, consisting of four broad lanes, each direction separated by a strip of
> grass, is magnificent. And bridges for people to cross over it. Spectators crowded
> on to these bridges and the sides of the road. A procession. And a glorious view
> as we were driving straight towards the Elbe and the Lössnitz Hills in the evening
> sun. We drove the whole stretch and back again (two times 7½ miles), and twice I
> risked a speed of 50mph. A great pleasure, but what a luxury, and how much sand
> in the eyes of the people. There are constantly accidents at hundreds of railway level
> crossings, thousands of roads are in the worst condition, everywhere there is a lack
> of cycle paths, which would do more to prevent accidents than all the tightening up
> of the law. None of it is done because of course it would not catch the eye. On the
> other hand "THE ROADS OF THE FÜHRER"! (4 October 1936)[31]

This passage again brings out Klemperer's infatuation with speed. Yet the pleasure
he is starting to get from being behind the wheel soon comes to grief because of a
string of accidents. On 22 May 1937, he collides with a cyclist, and on 25 November
1938, he loses control of the car while braking:

> The car rolls over the embankment and I'm suddenly lying on my back in the field
> right next to the car with a burning sensation in my face.

He and Eva sustain minor injuries, a doctor is called to the scene and the car has
to be towed from the field. Klemperer manages to drive it home with considerable
difficulty. (The outer rim of the steering wheel has been torn off, so Victor has to
steer with the bare spokes.) And that, in effect, is the end of the story. The couple
drive to Leipzig once more on 6 December. Shortly after, German Jews have their
licences collectively revoked; Klemperer utters hardly a word of protest, though he
notes with bitter sarcasm:

> The healthy sense of justice of every German manifested itself yesterday in
> a decree from Police Minister Himmler with immediate effect: Withdrawal of
> driving licence from all Jews. Justification: Because of the Grünspan murder

Jews are unreliable, are therefore not allowed to sit at the wheel, also their being permitted to drive offends the German traffic community, especially as they have presumptuously made use of the *Reich* motorways built by German workers' hands. This prohibition hits us terribly hard. It is now three years exactly since I learnt to drive, my driving licence is dated 26 January, 1936.[32]

To the movies – preferably twice a day

Klemperer and his wife were enthusiastic moviegoers – the diaries make it clear that they not only went often, but that they took their experiences seriously and thought through everything they saw. Victor was active as an arts journalist throughout his life – in the early Berlin years professionally, then episodically during his lecturing days – and made notes on what he read and saw from an early age. His diaries contain something in the order of 1,000 miniature film reviews and a wealth of observation that would supply the raw material for a remarkable account of the early cinema – a history told from the point of view of the consumer. I use the word *consumer* advisedly, for Klemperer himself was well aware that moviegoing was not just a continuation of middle-class cultural life in a new medium, but a powerful source of commercially generated pleasure as well, a form of relaxation in a virtual world:

> When the movie works, I feel good. Light music, operetta, opera, waltzes – they all make me feel emotionally alive, colourful scenes of adventure or daily life – magnificent looking women, men in tails and top hats – always something sexual, always a certain something else, a redemption from life's pressures. (10 October 1921)

It matters little what sort of pressures he faces – there is no phase of Klemperer's life that doesn't see him under pressure of some sort – the sense of emotional relief is always the same. In October 1921, he has again been overtaken by professional worries (and professional jealousies). At other times, he finds himself short of money. His struggles with writing or academic work are another constant. His entry for 28 September 1925 is characteristic here:

> We've literally been to the movies every day – either later in the evening or at 6. I'm in a tense state; pulling myself together for work is beyond me.

The same entry continues:

> I'm in a stale, confused state. The best thing for it would be going to the movies twice a day.

Alongside the sense we get from his 1921 diary that the movies bring him a sort of redemption, a snippet of life uncontaminated by real-life pressures, the comment that he seeks out something that will 'make him feel emotionally alive' is doubly interesting. Is the comment simply what one might expect from a middle-class academic who strives to lead as outwardly affectless a life as possible? In a sense, the same tendency makes itself felt throughout technological civilization generally, insofar as the general advance of technification brings with it an objectification of day-to-day existence and hence requires of us a relatively affectless stance towards the procedures and systems that determine the shape of material reality.

Klemperer's predicament, nonetheless, has something historically particular as well as something historically general about it. Looking to art for straightforward emotional gratification is something decidedly *improper* by the standards of the educated middle classes of Klemperer's day. Men and women of Klemperer's ilk thus either refused to see cinema as art or they saw moviegoing itself as something faintly salacious. This made things difficult for Victor and Eva. As a couple, they clearly aspired to live up to conventional bourgeois manners and mores from beginning to end. Victor, in particular, seems to have ardently wished for nothing more than the life of a respectable academic. Yet they were used to a more free-wheeling existence from their earlier years together, and the sense of never having firmly established themselves in middleclass society was to last from 1919 to 1935 – as long as Victor lectured in Dresden. That, in turn, meant that *they themselves* thought of their avid taste for cinema as *déclassé* and did their best to hide it from their official circle of respectable friends. Going to the movies was something embarrassing, as we realize when we read Klemperer's story of his encounters with a university colleague, Kowalewski. The two meet by chance, exchange rapid, rather frosty, greetings, and go their separate ways. Then, to their mutual surprise, they cross paths again not long after in a socially ambiguous zone rather close to a cinema:

> I tell Eva – Kowalewski doesn't need to know we're here for the movies, there are plenty of shops in the arcade [so let's simply avoid him]. In the meantime Kowalweski approaches with his little coterie of lady-friends, that nervy highly Germanic woman and the grey-haired Americanised one – either his sister or his sister-in-law. The ladies beam, Kowalewski is anxious. . . . But then after the film we took a little turn together as far as the *Altmarkt* and avidly confessed: what fun the movies are, how often we go, how annoying we find the aesthetic hypocrisy of the crowd that waxes indignant about film, how nervous we still are about going. (8 May 1921)

The middle class distaste for cinema is not just something they need to take into account socially, it is also something Victor and Eva have internalized. 'All in all, we were ashamed of ourselves', writes Victor on 17 January 1921, after they have seen an especially bad film.

We can leave aside Klemperer's reflections on cinema's aesthetic development, particularly as Nowojski allows himself plenty of editorial commentary about the masses of informal film reviews to be found in both volumes of the diaries. In the present context, we want to focus rather on technical developments, above all the introduction of slow motion and sound. Klemperer comes across slow motion for the first time in educational and propaganda films in the early 1920s. (Obviously, the technique was yet to be integrated into feature films.) The two relevant diary entries are for 17 July 1921, when he records his impressions of a documentary about football and gymnastics, and 13 April 1925, when he goes to see an edifying production called *Paths to Strength and Beauty*. On both occasions, he describes the use of slow motion as *instructive* or *informative*. His point is that slow motion makes it possible to get a much more exact sense of what is involved in physical movements than technically unmediated perception would normally allow.

On the earlier occasion, he also finds the effect of slow motion *funny*. This, presumably, is his first encounter with the new technique:

The player's head and the ball float towards each other – at this speed one doesn't experience the collision of the two *as* a collision – and then the man floats away again, sinking gently to the ground. (17 July 1921)

The comic effect is obviously the result of a shift in perspectives; the viewer's disappointed expectation of sudden action creates an emotional charge which releases itself in laughter. Here, as elsewhere, Klemperer's astute mixture of observation and self-observation shows how technically mediated perception produces a marked change in the way the viewer takes part emotionally in what he perceives.

In December of the same year, Klemperer sees his first sound film, *Atlantic* – the 1929 dramatization of the sinking of the Titanic:

What shook me though was a great innovation. "Atlantic" was the first major production to make full use of sound. (21 December 1929)

After detailing the film's basic contents, Klemperer comments on the impact of sound:

[Kortner's] voice sounds completely natural. The voices of others, particularly the female characters, still sound distorted, as though the actors had spoken their lines into a pot. The sound of running water comes across well, etc . . . . This film shook me, both as film and as tragedy. As a sound film I would say it is a major event, a watershed.

Sound films clearly take off from late in the decade, but their sound quality is so poor that they almost put the Klemperers off the movies altogether. Victor notes on 22 April 1930: 'Hardly ever going to the movies. (The curse of the sound film.)'. It is not until more than a year later that he sees a sound film that he thinks succeeds completely on a technical and aesthetic level, *Richard Tauber: Das lockende Ziel*. Curiously, it is in capturing *song* rather than the spoken word that 'talkies' first come truly into their own. Klemperer writes:

> Sound works (a) because it is a meaningful part of the film. Vocal play: the simple boy, the "high-German" folk-song sung by the artless youngster, the operatic aria, the opera-scene itself, the scene in the radio studio. And because film is also used *as film* [to show us] landscapes, open-air scenes, ever-changing locations; (b) because of the quality of the sound; admittedly everything sounded so good because almost everything was sung, almost nothing spoken. The possibilities of sound films are still strictly limited. Spoken dialogue is a hopeless mess. (30 June 1931)

Klemperer's 'still' in the second last sentence here is the marker of a quintessentially progressive attitude to technology. Rather than attributing the weaknesses of 'talkies' to technology *as such*, Klemperer ascribes them solely to the underdeveloped state of the technology of the day. Not long after, he declares himself satisfied with the sound quality of spoken dialogue too. The film that earns his praise is, in fact, the dubbed 1930 adaptation of Remarque's *All Quiet on the Western Front*:

> All of the dialogue came across perfectly, almost completely undistorted – which is even more curious, given how thick and fast it came and given that the German text had been inserted over the top of the English original. (26 July 1931)

However, not long afterwards, his opinion has swung around again:

> After several good experiences (with "The Million", "Richard Tauber", the Remarque adaptation), we wanted to give sound films another try. But we're shying away again after another excruciating experience. (15 August 1931)

It would be remiss not to mention the political dimension of film which Klemperer notices – and makes note of – just as he had in recounting his driving experiences. Time and again, he finds himself willingly or unwillingly confronted with propaganda films – and the politically charged reactions of audiences:

> Time and again what strikes me is the politicization of the audience. Every time the Imperial Army of old or a German flag or "our" Hindenburg appears

on screen there's wild applause. The same sort of positive publicity for Stresemann and Luther's Locarno peace deal got hardly any response at all. (29 October 1929)

Conclusion

The period that opens out to us in Klemperer's diaries from 1919 to 1938 is the era known as the interwar years. In the eyes of a sensitive contemporary observer such as Klemperer, they are years of dynamic transformation suffused with expectations of an equally dynamic future. Technological innovations reach into the very midst of daily life. And Klemperer and his wife absorb them in an open-minded spirit, sometimes with enthusiasm, interpreting them as emblems of freedom and as points of departure for future developments. In the 1930s, the same dynamism becomes linked to the emergence of National Socialism. The *Volkswagen* – literally the 'people's car' – was soon to become the patriotic symbol of the synthesis of German nationalism and technological progress.

The new products and the new media of the period were, on the whole, not refinements of existing technologies; they turned up in the lifeworlds of men and women like Victor and Eva Klemperer almost totally unexpectedly. Apart from the electric stove, the new technologies of the 1920s and 1930s brought with them new forms of life and new life experiences. They changed everyday existence in major ways: with the car came not just a new form of leisure, the road trip, but a new individualizing sense of mobility; the cinema was not just an expanded version of that paradigmatic institution of educated middle-class life, the theatre, but the vehicle of a shift in the possibilities of perception; while the gramophone was likewise the cornerstone of a new form of sociability.

The powerful sense that the diaries give of the emotional engagement that accompanies the introduction of each of the new technologies is extraordinary in many ways. The easy acceptance of the new forms of technology, the pleasure they gave, the enthusiasm and the disappointments they caused, were clearly a vital dimension of the life of a generation. That makes the testimony of a man like Klemperer all the more remarkable, for, as the diaries make abundantly clear, he was both without a practical affinity for the new technological forms and completely without a professional knowledge base on which to build an interest.

4

The technification of human relations

Technostructures: Society and nature

Definitions

When confronting a subject as complex as the relationship between technology, society and nature, the first problem we face is the understandable demand that we define the basic terms of the discussion. After all, it is this expectation, this urge to define, that first brought philosophy into the world, this expectation that philosophy has never been able to satisfy. Why not? The answer has something to do with the interconnections between the concepts, in this case the fact that technology, nature and society are so intertwined that stipulating in advance the meaning of each individual term is a hopeless task.

In this chapter, however, we want to speak to the need for a basic first approach to our problem with an historical reflection. It is this: the conceptual difference between technology, society and nature reflects an historical differentiation between the three, which in practice has been overtaken, and indeed rendered obsolete, by the course of modern developments. In his *Physics*, Aristotle makes a distinction between natural and technological objects; he tells us that the first contain their own principle of movement while the second do not.[1] The famous example he uses to point at the difference is a bedframe made of willow: if one buries the frame, says Aristotle, what will grow out of it will be another willow tree, not another bedframe. One sees immediately that *man* is the unspoken third term through which a distinction between nature and technology first appears: the production/reproduction of all things technological requires the input of human agents; nature, on the other hand, reproduces itself. Human beings, insofar as they are *socially and culturally organized beings*, are precipitated out of the two-way solution, for though, in a certain sense, man possesses his own principle of movement (i.e. re/produces himself on his own initiative, i.e. naturally), his methods of re/producing himself also regularly demand the sort of activity that would qualify precisely as technological rather than natural: in general, man reproduces himself and the broad fabric of his life in a *deliberate* way through *rationally organized procedures*, above all through the production of material goods. The self-reproduction of man as a social being does not simply take place by itself; it is a socially organized process, rather than a natural process.

Thus, according to the classical division of the ontological ground, technology is something other than nature, nature is something other than technology and social man is neither of the two. And yet if we expand our historical frame of reference, it becomes obvious that these classical distinctions are today quite untenable. The natural environment that surrounds us is simply no longer capable of reproducing itself without human assistance, while technology and society seem so much a part of one another that it is possible, indeed entirely sensible, to talk nowadays of the *self-reproduction of technology*; humankind, we could say disparagingly, is as it were the DNA of the virus we call technology. These are anticipations of the argument that we want to run. Our task in what follows is to sketch out the detailed connections that exist in the triangular relationship between technology, nature and society.

The technostructure of society

Let us begin, however, by approaching our central problem, the fundamentally social character of technology, from an oblique angle, viz. by looking at the fundamentally technological structure of present-day society. In this context, we can bring together some of the results of our analysis in Chapter 2.

One of the main reasons why the social character of technology repeatedly eludes our grasp is that the blanket term 'technology' calls up a picture of *individual* technological objects or *individual* technological processes. And yet our social condition, insofar as it is shaped by technology, depends less on separate technical things or procedures, and far more on the existence of technological *infrastructure*. As we have already seen, it is not telephones or mobile phones but the systematically interconnected telephone network that is constitutive of the social meaning of technology – not individual trains, but the entire rail network, not cars, but a transport system designed for individual passenger travel made up of streets, petrol stations, oil refineries, car factories, panel beaters, insurance policies and so on. The way that contemporary society makes use of technology is *not* primarily as an individual means to particular ends; rather, contemporary society itself has a technologically determined internal structure, a technological *infrastructure*. The consequences of this, moreover, are momentous: the first is that social self-reproduction brings with it the necessity of reproducing a technological infrastructure; the second is that social development inevitably involves the continuous further development of that technical infrastructure. It becomes clear why technological development has come to constitute a special political field, and, indeed, why the development and introduction of particular forms of technology tend to be beset by controversy[2]: in short, because such issues relate to utterly basic structural elements of the societies we live in. The systematic organization of energy production and supply, together with the systematic integration of data infrastructure networks, constitute the skeleton, or perhaps we should say the armour plating, of our entire social organism.

The various technical systems of material means, especially in the developed industrial world, have come to make up a massive social superstructure – the third of those collective creations of society which, once in existence, seem to stand strangely over and against human beings and operate according to a peculiar logic of their own. The first two of these structures are *the state* and *the free market economy*, which are being joined in today's world by the *technostructure of society* as a third major element of social being. From the point of view of the individual's place in society, we might say that the historical picture looks roughly like this: in the estate-based states of the early modern period, with their strata of noblemen, clergy and commoners, the individual's membership of the social whole was largely determined by his place in the organism of the state. In the bourgeois societies of the eighteenth and nineteenth centuries, the individual belonged to the social whole insofar as he was part of the division of labour, as a participant in the market-based exchange of goods and labour. In the current era, the individual's membership of the social whole comes to be determined by an additional factor – his role as a participant in the total system of technological infrastructure, as the holder of a position within our society's all-pervasive technical networks. In the latter case, however, what at first sight may appear to be a *merely* technical form of social integration tends to carry a political charge as well.

In short, it seems important to recognize (1) that social reproduction includes the reproduction of a vast technological infrastructure; (2) that technological integration is one of the major aspects of social integration in today's world; and (3) that the organization of labour and the possible forms that labour can take both closely reflect existing forms of technology.

This third point might, at first sight, appear too trivial to mention; it has certainly been clear for a long time that an expansion of machine production would, in the long run, require less manual labour and more technical expertise, just as it has become clear in more recent times that computers have made a large amount of middle management redundant. Yet it remains necessary to mention this third point, and indeed to underscore it, because of a misleading assumption that is common both to public debate and theoretical discussion, viz. that technological innovation takes place *essentially* within an *isolated* technical dimension of social life – that its effects within the labour market are, in essence, side effects. Our view is the opposite: technological action, technological innovation, etc., are for us *thoroughly* social forms of action. Technological innovators aim to rationalize workforces and increase labour productivity; dramatic reductions in the necessary amount of *human* labour involved are essential to realizing their goal, not an unfortunate by-product whose consequences we should simply try to mitigate after the fact.

As we have argued in Chapter 2, technology today is no longer an arsenal of material means that we can make use of to reach certain ends. We live today in a technological civilization, and what that means is that numerous aspects of social life, indeed of life *tout court*, have become technological processes. Neither mobile phones nor their immobile predecessors are simply the means for carrying on conversations. The telephone, on the contrary, is a form of communication unto itself; its widespread use changes the very meaning of conversation. Jet aircraft, similarly, are not a technological aid to high-speed travel that one makes occasional, or, increasingly, everyday use of; they too change the meaning of travel, making it into a thoroughly technified process. Our examples are relatively simple and they might be multiplied. What they show is that it is misguided to think of technology as something external to human life proper – as something beyond which what is *truly human* has as much scope to flourish as ever. On the contrary, as we will see in more detail in this chapter, a large proportion of social life today is of a *thoroughly* technified nature: our relationships with our fellow human beings, and indeed our relationship to ourselves, are frequently defined by technological action.

The social constitution of technological products

Our first approach to the relationship between society and technology has shown that technological action is *always already* social action because each and every technological change, each act of reproduction or renewal that takes place within the system of material means, at the same time serves to reinforce the technostructure of society. The technostructure is, as it were, the skeleton of social life, governing forms of communication, the division of labour, the individual's membership of society and a variety of forms of domination. From the technological constitution of social relations, we turn next to the social constitution of technology itself. The question which is of prime interest here is the question of the social character of individual technical products, or, more specifically, the question of how far society (social values, social expectations and social forms of organization) makes a difference to the material structure of technological objects.

It has always been tempting to interpret technology as applied natural science, and though that view of the matter might contain some truth, it is also mistaken in several essential respects. Taken by itself, the idea that technology is simply applied science turns out to be quite false. Obviously, every technical process, every operation that every technical device performs, every regularity exhibited by technology in action, is specifiable more or less directly in terms of the laws of nature. Yet however well founded that presupposition is, it masks the fact, firstly, that in practice technological results are, as a rule, in no simple sense derived from the laws of nature, and, second, that in the vast majority of cases it would be next to impossible to mathematically deduce the workings of technological devices from the laws of nature underpinning them. All that the theoretical presupposition that technological processes occur in accordance with the laws of nature means in practice is that nature is assumed to play its part – play along as it were – in the creation of technological objects and processes. Whether nature does indeed play along is usually something that is verified practically, i.e. empirically. Scientists will indeed attempt to compress the process of verification using various models and/or simulations; however, there can hardly ever be a question of directly deducing the behaviour of the technological devices one is designing from the actual laws of nature. Construction, rather than deduction, is the fundamental driver of technological invention.

Engineers are thus not really a species of the genus scientist whose task it is to specify the detailed workings of the natural world after science has uncovered their broad outline. Instead, engineers design new processes and movements within the broad framework of what is possible in accordance with the laws of nature. That, in turn, raises the question of what technically constructed processes and movements actually *are* and how they relate to natural processes. The oldest answer, which we owe to the traditional arts and crafts, revolves around the concepts of form and matter, or form and content: the idea, in short, is that the work of human beings stamps natural material with a form of one kind or another, and that that form is essentially an expression of human purposes. Of course, the form/matter distinction is, in many ways, outdated: on the one hand, the technological effects that engineers and inventors contrive are not simply a matter of static forms imposed on pre-existent material, for what technologists give shape to are processes, or even systems. On the other hand, the material that goes into the formation process is not generally taken over wholesale from nature, but almost always something that is itself either arranged (preformed) by human beings or constructed by human beings from scratch. Nonetheless, the form/matter model still contains a grain of truth insofar as what technology brings into being is always a network of human purposes – a network of social purposes. We could say essentially that the function of engineers in relation to nature is an organizing function. Engineers, in short, arrange nature and its elementary building blocks for social purposes, though that is equally true of everyone from the traditional craftsman to large teams of modern-day technical systems builders.

That way of putting it leaves open what nature's basic building blocks and what social purposes actually are – a point it is wise to remain unspecific about if we are to take into account both the historical variability of human beings' technological labours and the associated historical change in the natural elements technology makes use of. In his *Human History of Nature*, Serge Moscovici has attempted to chart this rough historical trajectory by breaking it into three stages or 'conditions', in each of which the basic form of human labour correlates with a basic conception of nature.[3] Moscovici designates his first historical condition as *organic*: this is the stage dominated by manual or craft labour in which nature is seen as raw physical material standing opposite the formative work of the craftsman. Moscovici's calls his second condition the *mechanical* condition. At this stage, nature is equated with physical energy and stands opposite the work of the engineer, whose task consists of the conversion and transmission of different forms of power. The third condition, which we are progressing towards in the contemporary world, is something Moscovici calls *synthetic*: human labour at this stage comes (is coming) to consist of the conscious design and careful regulation of nature considered as a system thoroughly amenable to human purposes. One of Moscovici's essential points at all three stages of his historical study is that we arrange and rearrange the natural world for our own fundamentally social ends.[4]

Moscovici is not the only author to put forward a persuasive argument to that conclusion though. Sybille Krämer has also made the point well in her work *Technology, Society and Nature*, where she writes of technology's capacity to *model social functions*.[5] If we are to usefully follow Krämer's lead, however, then we have to circumvent the problem she causes herself by reading technology as a system of means defined largely by the end of *economic* production. *Pace* Krämer, it is important to note that technology has been caught up in a range of different social contexts, more or less related to economic production, since the beginning of human history; everything from warfare to sport and play, from cultic activity to transport and traffic, from communication to political representation, has played a role in determining the course of technological development.[6] The decisive point here is that for new technological forms to materialize, the way a given goal is reached needs to be *socially stereotyped*; the crucial step is when a means or a set of means are detached from the particular situations in which they functioned solely as the means for particular individuals to achieve purely individual ends. Human forms of action, such as travel or killing or interpersonal speech, have to first take on a *general* cast in order to then be *recast* along technological lines. (This is not to say, as we have repeatedly stressed, that the material reconstruction of social functions takes man as its model insofar as he himself exercises such functions as an organic being. Car travel is not a mechanical reconstruction of human walking, though cars certainly represent an arrangement of natural elements for the same purpose as walking, viz. getting from A to B.)

So, in summary, (1). Though the physical events that technology sets in motion take place according to laws which are themselves part of the more comprehensive laws of nature, they are also, fundamentally, a realization of social functions. (2) In the day-to-day practice of engineers, it becomes impossible to overlook that every technological solution to a problem is the solution to a problem posed by human social co-existence; every optimization of a technical variable is the expression of a *social value*. The practical labour of finding technical solutions to problems brings out the social character of the technical solutions particularly clearly, for there are always multiple paths that lead to each particular concrete goal. To choose a path based on considerations of time effectiveness or cost effectiveness, as engineers do on a daily basis, means mirroring the way that human beings organize their common lives in systematic social terms.

The social values which confront engineers in their work in the form of numerous explicit rules, norms and laws are something we have summarized elsewhere under the heading of the *normative framework of technical action*.[7] Specifically, what this normative framework consists of are the officially recognized rules governing technology, most of which are detailed in a comprehensive body of regulation.[8] Other laws and statutes govern the environmental or public health standards that technological products are required to meet.[9] With the normative framework of technical action in the picture, we can now improve on the thesis we started by examining, the idea that technology is a form of applied natural science. Technology, we might say, is in our view more of a sort of balancing act between nature and society; in broad terms, the overall enterprise we call engineering is a search for solutions to problems that satisfy the laws of nature *and* the norms of society in equal measure.

To complete this section on the social character of technological products – a few comments about the structure of technological *knowledge*. Clearly, it follows from what we have argued that the greater part of engineers' knowledge is *not* pure scientific knowledge but *social* knowledge. But what exactly does that mean? When we speak of knowledge, it is important to make a distinction between two basic types: first, a form of knowledge that is inseparable from individual human beings (from an individual *person*); and, second, a form that is fully objectifiable, i.e. knowledge that is, in principle, accessible to everyone. In historical terms, it was precisely technological knowledge that started off as a matter of *experience*; and so it in part remains – technical knowledge is still a kind of *competence* in dealing with certain kinds of things and situations which the individual acquires through long practice and is, as such, not directly transferable to others. However, that historical state of affairs was to change considerably with the advent of modern science, one of whose characteristic marks is to detach knowledge from the individual personality. Parallel with the development of modern science, what we see is a bracketing of personal (and also local) influences in the field of technological endeavour. Technological solutions to problems are reduced to elementary forms, broken down into simple units that can be applied in many different situations. As solutions are standardized, they are incorporated into a canon of technical expertise. The consequences for the individual practising engineer are major; faced with a concrete problem, he has no need to take each step towards a solution alone, but can draw on a set of tried and trusted solutions, using the general stock of basic knowledge as a foundation for a more or less novel individual approach. Alongside the pressure to rationalize and provide a maximum of security, one of the main drivers of technical standardization is making standard solutions to problems widely accessible and easy to communicate. Technical knowledge becomes social knowledge in the process. Gradually, what comes into being is a social stock of knowledge which the individual helps to build up. Technical progress becomes a collective enterprise. As in science, every individual technological innovation is seen increasingly as a contribution to the general stock of technological knowledge.

Technological appropriation of nature

From the outset, we have tried to locate our reflections in relation to a trio of concepts, viz. technology, society and nature. In doing so, it has been important to bear in mind that, historically, a conceptual opposition existed between technology and nature. Natural objects are not technological objects and technological objects are definitionally non-natural, or such at least was the historical Western view anchored in Greek philosophy. However, with the growth of modern science, the Greek opposition of the natural and the technological was gradually set aside, and in a quite specific sense. In classical antiquity and in the Middle Ages, mechanics was not part of natural science. Its purpose was to wrest from nature by *cunning* what ran contrary to nature's ways (cunning indeed is the literal meaning of mechanics – *mechanike* – in the original Greek sense). Modern science essentially begins with the insight that what artificially contrived processes make possible is as much a part of nature as anything else. The methodological consequences of overturning the Greek view are dramatic: from Bacon onwards, scientists came to believe that the activity of the natural world could only be known in the full sense if it was brought to light unambiguously and under repeatable artificial conditions. Mechanics had become the prototypical modern science, and in many senses the secret of the applicability of scientific knowledge to the real world can be explained in terms of this initial interpretative posit. The knowledge of nature that science opens up is a knowledge of nature that *already conforms to technological conditions*. We can say that the material appropriation of nature, either by means of sophisticated craft tools or by more advanced technology, is a fundamental presupposition of natural science. To arrive at the knowledge of nature in this natural scientific sense, one must already be in a position to technologically control it by constructing high-precision scientific apparatus, by altering parameters and borderline conditions at will and by fabricating pure materials for use in experiments.

Yet if recognizing the historical affinity between technology and natural science is the positive result of our analysis, the negative implication is that neither technology nor natural science take *external nature itself* as a conscious point of reference at all. Scientific and technological knowledge is essentially only knowledge of a kind of sanitized, home-grown nature. Moreover, the limitations implicit in the beginnings of the modern scientific project become increasingly explicit as the project unfolds. The blindness of modern technology and natural science towards nature in the wider sense has come back to haunt us, for in today's world it is becoming painfully obvious that external nature – the excluded aspect of the modern technoscientific project – has essentially been treated as a bottomless reservoir equipped with unlimited powers of self-regeneration and an inexhaustible capacity for self-reproduction. As the immediate consequences and the plethora of side effects of technological development make themselves felt, we are forced to register that these cherished assumptions about nature no longer hold, just as we are forced to reassess our conception of technology, the so-called side effects of which we now realize must be consciously taken into account as though they were the direct goals of action.

We are beginning to realize that our restless technological activity depends on nature's capacity to play along; in abstract terms, we are forced to acknowledge a kind of spontaneous self-activity of nature. The possibility that capital-N Nature could again become a scientific theme as it was during classical antiquity seems to beckon; we again begin to see nature simply as *that which occurs of its own accord*, no longer as an objective complex of isolable states of affair that we come to know under technical conditions. Yet the clock can hardly be wound back. At precisely the moment we urgently need to acquaint ourselves with nature under something other than laboratory conditions – that is, just when we are on the verge of (re-)apprehending nature without our modern technological preconceptions – we come up against the sad fact that external nature no longer exists of its own accord as it might once have done. Under the imprint of technological activity, nature itself would seem to have become a social product.

In arguing that external nature has become a social product, I am of course not referring to the stars or the interior of the earth, but to *environmentally relevant nature* – the surface of the earth and the surrounding atmosphere or what we might call *medium-scale nature*. Nature on this intermediate scale is no longer even properly describable without specifying what sorts of causal interactions with human beings it is enmeshed in. Human technologies of production, human forms of transport, the structure of economies and the legal regulation of human behaviour vis-à-vis the environment are all major factors that shape the condition of the natural world around us – a realization that needs to be echoed at all levels of human knowledge, including at the level of a truly *social natural science*,[10] i.e. a science of nature which has taken on board a portion of social science, as natural science must if it is to adequately comprehend the natural/social complex that is its true object. One can only hope that such a socially inflected science of nature will soon be given the chance to begin its task of thought, for its practical relevance would be beyond doubt. The environmental impasse that global civilization seems to have reached does not demand knowledge of nature *per se*, but knowledge of the *socially constituted world of contemporary nature*, above all because it is knowledge of socially constituted nature,[11] not conventional natural scientific knowledge, that is the fundamental precondition of an environmentally aware use of technology.

Technology will continue to provide socially determined forms for appropriating nature for as long as human craft, human technology and human science continue to exist. The problem is that, until now, all that has counted in science is a nature that had already been appropriated – the actions and reactions of a nature that was already part of technological systems.

Today, what we need to make allowances for is that all technological action is also, in a primary sense, a form of action vis-à-vis external nature. Large-scale technological projects transform whole landscapes and play havoc with entire ecological cycles. Natural resources are depleted and materials found in concentrated deposits in the earth's crust are dissipated. So it would seem that we have to realize that technological activity is not just an *appropriation* of nature; like all forms of natural metabolism, it also involves the excretion or elimination of the nature that we use up as well. Above all, we have to be clear that each and every technological process is also a natural process with a host of effects and that the distinction between the primary effects and the side effects only has a meaning in the context of social values of our own choosing.

Every technological undertaking is a form of action in external nature and upon external nature. Our argument is that we delude ourselves if we take technological action as something less than action in the full sense of the word, or treat the impacts of technological action on external nature as the *incidental* side effects of an activity that is *essentially* directed to other goals (production, transportation, communication, etc.). The latter view, at best, leads to environmental protection becoming one of a host of considerations in the design and execution of technological projects. And, indeed, environmental protection is something that environmentalists are right to insist on, for it is clear that the regenerative powers of external nature are still great and that we should take care not to place any further strain on them. Yet environmental protection is still no more than a noble form of rearguard action; the environmental resources it sets out to protect dwindle noticeably by the year. If, by contrast, we take the idea of a socially constituted external nature seriously, the actions of human beings themselves might show up as a prime factor in the reproduction of external nature; we might at last begin to face up fully to the fact that the world of nature is incapable of reproducing itself to anything like the degree we wish it to without the conscious intervention of human beings.

What has been happening simply *by chance* over the course of centuries and millennia, viz. the transformation of external nature by human action, stands before us today as an inescapable *duty*. Our environmental/historical predicament demands that we make the condition of external nature a conscious object of forethought and care. If that is so, then every major technological undertaking has to fit into the framework of a general politics of nature. The way technological projects will contribute, for better or worse, to *the entire future world of external nature* has to be considered fully for the first time, as does the explicit human labour that technological projects will make necessary if they are to remain compatible with the reproduction of external nature. The fundamentally social character of the natural world around us thus means, at the most basic level, that every technological undertaking has to be regarded as a conscious act within an overall politics of nature.

Summary

The results of our analysis of the triangular relationship between society, technology and nature can be summarized in the form of three theses.

We live in a technological civilization, which means that we can no longer adequately understand technology from the point of view of individual technological devices and their individual purposes. Rather, our society has acquired an all-pervasive technological *infrastructure*, which has become one of the major forces that shape the very possibilities and meanings of social life. The introduction of new forms of technology is thus part of the overall process of social reproduction, and brings with it inevitable changes to the very structure of our society.

Technologically mediated events are not simply processes that take place in accordance with the laws of nature, but are arrangements of natural forces and elements governed by social requirements. Technologically mediated events give external or material form to functions that are deemed necessary by society. The social character of technological objects and devices is not something they acquire, as it were, retrospectively by being used, but something deeply embedded in their material structure itself.

The world of external nature must be seen as socially constituted. Technological forms of activity should not merely take into consideration their potential effects on external nature, they should be regarded as conscious attempts to shape external nature. The design, deployment and use of technology, in short, must be seen as conscious political acts and must be assessed as part of an explicit politics of nature.

Anthropological change in a technological world

The most general task of philosophical anthropology is to explain how human beings understand themselves *as human beings*. The task presents itself forever anew, and so we could say that the conclusions of philosophical anthropology will turn out differently time and again, depending on the society and the historical context in which a particular mode of human self-understanding has come into existence. Perhaps that is not as obvious as might appear at first sight. After all, is it really the case that human beings are forever reinventing themselves? Anthropology has been understood by some as the very opposite of history,[12] its fundamental concern taken to be what have been called anthropological universals, those invariable aspects of human life that determine our existence transculturally and independently of history.

My own work in the field of philosophical anthropology took shape in response to a more specific question, viz. what it means to be a human being in a world whose self-image has been fundamentally influenced by the so-called sciences of man. The very existence of a type of scientific knowledge that takes human beings as its theme has profound consequences for the way we live as human beings. If the sciences of man treat human beings as a kind of foreign object, then the task of philosophical anthropology is to explicitly pose the question of what it means for human beings to become objects of knowledge. Since the goal of my own work was to draw practical conclusions about the scientific knowledge of man, it seemed fitting to give it the Kantian title, *Anthropology from a Pragmatic Point of View*.[13] In the present essay, I am going to focus on a question that is related to the major preoccupation of that earlier work: what does human existence mean in a technological civilization? A handful of theoretical concepts that belong in the context of my first attempt at philosophical anthropology have thus been taken up in my analysis here.

The notion of anthropological change within a technological civilization only begins to make sense, it must be said, if one assumes that the changes underway in the technologically driven world around us are *deep-seated* changes. But what does it mean for a change to be truly deep-seated? One answer could be that deep-seated change is precisely the sort of change that affects the very *essence* of man. This answer goes against my position in *Anthropology from a Pragmatic Point of View*, where I set out from the existentialist premiss that man as such has no essence. The formula commonly attributed to existentialism speaks of the precedence of existence over essence. Heidegger alludes implicitly but clearly to the idea, for instance, when he claims that 'we cannot define Dasein's [i.e. man's] essence by citing a "what" of the kind that pertains to a subject-matter [because] its essence lies rather in the fact that in each case it has its Being to be, and has it as its own',[14] while Sartre takes the idea one step further: 'What do we mean by saying that existence precedes essence? We mean that man first of all exists, encounters himself, surges up in the world – and defines himself afterwards. If man as the existentialist sees him is not definable, it is because to begin with he is nothing. He will not be anything until later, and then he will be what he makes of himself. Thus there is no human nature'.[15] If the being (existence) of man is what defines his essence, for Sartre at least, what that means is merely that the *individual*'s existential self-definition is all important. This, however, is where philosophical anthropology and existentialism part company, for, from an anthropological point of view, there is no escaping the fact that historical, social and cultural conditions all define quite different forms of human existence. The thesis that there are alternative forms of existence that are *variously* individual, historical, social and cultural has been given plentiful proof since Heidegger and Sartre in the fields of historical anthropology[16] and ethnology. Empirically, the conclusion is well supported: what it means to be a human being is something that has a quite different structure from one time or place to another.

Instead of talking about an essence of man, my preference is to talk about *anthropological conditions*. The latter, we might say, are different ways of organizing various *humana*, where by *humana* I understand certain fundamental possible modes of determining what human existence means – body, soul and spirit, for example, or language, work and historicity. It is important to insist that these *humana* are *not*, collectively, a substitute for what was once known as human essence or human nature; on the one hand, because the *humana* take on ever new forms throughout human history (thus we can speak, for example, with Bruno Snell of the *discovery of the spirit*[17] or with Hermann Schmitz of the *invention of the soul*[18]), and, on the other hand, because they stand in mutual relationships of expression and repression. The last theoretical concept that is implied by our way of looking at things is thus the concept of *difference*, which allows us to articulate the fact that the cultivation of particular *humana* – the way different *humana* are stylized in different historical periods and cultures into different conceptions of the essence of man – necessarily renders other *humana* obsolete, indefinite or unrecognizable. One well-known example of *difference* in this sense would be the self-stylization of Enlightenment man as a rational being – the cultural effect of which was to reduce the life of the human body to mere irrationality, or, indeed, to bestiality.

In theoretical terms, anthropological conditions are historically and culturally conditioned meaning-giving structures within a wider, never fully delimited field of *humana*, while the *humana* themselves are the building blocks of human self-stylization in terms of one or another human essence, as well as the basic means of excluding various aspects of human existence from a given conception of human essence. Such a theoretical notion of anthropological conditions, we note in passing, is explicitly intended to allow philosophical anthropology to step beyond its customary Eurocentrism; not by presenting the terms of our analysis as a universal evaluative standpoint, but by taking almost the opposite tack, viz. by overtly practising philosophical anthropology as a process of self-understanding. In short, we need to bear in mind that our contemporary anthropology is itself the expression of European intellectualism within a technological civilization. Using other non-European modes of human existence as points of comparison and, in particular, ruling out all forms of teleology which posit European intellectualism as the ultimate goal of human history is absolutely crucial.

As a full discussion would take us too far afield, I will outline my view of anthropological change within a technological civilization in the form of four theses[19]:

Thesis 1: The growing importance of technology within contemporary lifeworlds has brought with it a reversal of the process of civilization described by Norbert Elias. The past 100 years have certainly not seen any increase in human beings' average level of inhibition; the broad social trend would seem to be towards a marked tailing off of our general sense of shame. Similarly, the process of transforming external compulsions into intrapsychic compulsions – another of Elias's characteristic marks of advancing civilization – has swung into reverse, and while, on the whole, direct physical or political violence has come to shape individual behaviour less and less (in the developed world at least), the brute force of the material world, and in particular the technification of the lived worlds of work and transport, has more than made up the difference. Because sanctions against disobedience or error in technified lifeworlds follow almost automatically, an internalized moral framework to keep individual waywardness in check becomes increasingly superfluous. Since the early twentieth century, we can observe a further result of this intensification of external pressures in the form of a gradual breakdown of intrapsychic mechanisms of self-control.

Thesis 2: The system of material means supplied by our society's vast technological superstructure has brought with it a far-reaching separation between instrumentally rational, goal-directed action and the pursuit of everyday life fulfilment. Looking back from our contemporary situation, where we find purposive activity and life fulfilment largely split off from one another, it might almost seem that we at last have conclusive proof that nature pursues its aims indirectly or even haphazardly. By that, I mean that what human instinct aims at is, in general, something different from any straightforward natural purpose. Our thesis, however, proposes that the distinction between an instinctual aim and a natural purpose, which formerly only made sense analytically, now comes to have full practical validity, as two types of action diverge from one another ever more dramatically under the influence of technification. One prime example would be the relationship between sexual pleasure and procreation. Similar developments include the separation of eating and physical nourishment, movement and traffic, or thought and calculation.[20] To be sure, the practical division between each of the paired terms is far from complete, and again we need to add a dash of science fiction to see where current social tendencies might lead: imagine, for instance, if the reproduction of the human species were to become an ever more technified business while the hyper-sexualization of human social life were to continue unabated; or imagine that physical nourishment were to become a matter of medically stipulated doses of nutrients and vitamins, while at the same time our present-day eating rituals – our entire cult of fine cuisine – were to become ever more sophisticated and elaborate.

Thesis 3: Technological civilization has led to the emergence of an increasingly independent world of images which absorbs ever more attention and energy, and is governed increasingly by its own rules. However, in order to understand what that means, we would first have to clarify what the general relation between images and emotions is.[21] Let us briefly propose the provisional thesis that forming an image of reality once literally involved the *formative* action of the imagination, i.e. bringing the body into dispositional relation to reality through a process of mental figuration.[22] Genealogically speaking, imagery would have had the function of activating human beings' potential for bodily excitation. However such a genealogical hypothesis shows contemporary technological civilization in an extremely curious light, for what would seem to be taking place in the contemporary world is, as it were, a double movement away from our (hypothetical) original condition as the worlds of fact and fiction move off in different directions and, in particular, as the world of images acquires ever more independent reality: on the one hand, we see the organized impoverishment of experience in the name of security and control (which amount to nothing less than the highest maxims of our technified world) and, on the other hand, we see the deployment of manifold technical means to provide a substitute satisfaction for the very emotional energies that technification has rendered useless. With the rise of the novel, then television, and now a profusely imaginative new generation of computer games, an increasingly large part of human life takes place in an anonymous zone composed of images that are increasingly disconnected from reality.

Thesis 4: Technological civilization has seen the integration of all manner of esoterica into exoteric everyday life. Though originally practised within elite circles of monks and other adepts, mystical religion, meditation and yoga are entering into contemporary mass culture, usually in highly truncated forms. Contrary to the beliefs of some of their advocates, however, the process is by no means a countercurrent to the broad stream of technified life; if anything, it points up something quite essential to a technological civilization. As the body becomes more and more dispensable as part of social labour, a certain sort of discovery, or rediscovery, of the body becomes possible. As physical presence becomes less and less relevant to social action, bodies become increasingly free as vessels of individual pleasure. And as many kinds of rational thought, particularly of the repetitive and calculative variety, are delegated to machines, human beings are less and less restricted to linear, object-based thinking and are freed to explore other forms of consciousness. The broad anthropological condition that we could call Cartesian dualism – the definition of human existence in terms of a faculty of understanding/reason and a body considered as a machine – is something that a technological civilization gradually renders problematic. In doing so, it makes room for new anthropological structures and new forms of human existence.

The technification of perception

Structures of perception

The relationship between technology and perception is by no means a particularly novel object of theoretical interest; indeed, discussions of the relationship have almost come to define the philosophy and history of technology. In particular, perception is the source of a certain sort of example that is adduced time and again to explain technology's very essence. The question of what technology in essence *is* has often been taken to imply a series of further questions about the meaning of technology for human life (the anthropological dimension of our problem) and about its meaning for society (the sociological and economic dimension). Yet *what human beings and human society themselves are* have normally been taken to be fairly straightforward; a host of simple assumptions about each have produced the further simple assumption that technology is a means of achieving certain humane ends, or a means of fulfilling certain social functions. If, in this essay, we have chosen to reformulate the question of technology and perception as the question of the technification *of* perception, then we do so in the belief that the concept of instrumental means does a poor job of capturing either the human or the wider social meaning of the technological enterprise. More precisely, my argument will be that we find ourselves caught up today in a phase of human history in which technology is encroaching on what it is to be a human being and what it is to have a human society in a deep and radical way. Just as we have dealt in previous chapters with the technostructures of contemporary society, so in the present essay we will focus on a number of characteristic cases in which perception itself has come to be structured along technical lines.

We live in the midst of a technological civilization – could anything be more obvious than that? What people who make that claim generally have in mind is the ubiquity of technology, the indispensability of the technological devices and technological infrastructure at our disposal. And yet it is this very ubiquity and indispensability which means that we take technology for granted in a certain sense – which means that it becomes curiously *inconspicuous*. The extent to which it already influences the formation of individual personality or the general form of human self-understanding is very difficult to pin down; its effects on social institutions are just as diffuse as its influence on patterns of behaviour, modes of perception and forms of work and movement. Altogether, we seem to lack a cultural history of technology. However, we do possess numerous studies of the introduction of particular technologies, as well as different types of comparison of technified forms of behaviour and technified social formations with their non-technical equivalents; what much of that research shows is that developments at the interface between technology and perception do not merely play themselves out in the present, but in part stretch back to distant historical periods that it would be highly counterintuitive to describe in terms of any notion of technological civilization.

Let me immediately underline what has just been said with the help of two short excerpts from the sort of studies that I have in mind.

The first, which comes from Paul Feyerabend's *Against Method*, relates to the introduction of the telescope as an instrument of scientific observation at the time of Galileo. In *Against Method*, Feyerabend mentions that his research has pursued the hypothesis that 'The practice of telescopic observation and a knowledge of the new accounts of the heavens did not just change what one saw through the telescope but also what one saw with the naked eye'.[23] We note immediately that the theory that Feyerabend is careful to label a mere hypothesis might indeed seem an extreme one; the claim being tentatively considered is *not just* whether the scientific use of the telescope brings about an adjustment of human vision to the demands of a technical device, but whether the *mere existence* of the technical device changes the very structure, what we might call the culture, of vision.

My second passage comes from the work of the historian of the body, Barbara Duden. It relates to techniques of visualization in the field of reproductive medicine, and in particular to the creation of artificial images of the child in the womb using ultrasound technology. Duden draws a clear line between the types of questions she wants to pose of such technology and the types of questions one might expect from a more conventional assessment, focusing squarely on direct effects. On Duden's account, the latter type of inquiry brings to the fore issues like 'what medical options does the use of the technology open up? What effects will it have on the legal position of women? What damage could ultrasound do to body tissue? Whom does ultrasound technology ultimately exist to serve and how does it do so? What are the economic consequences of its use in Germany and the wider world? What are the alternatives to the use of such costly techniques? How does the use of such technology introduce qualitatively new stages into the course of pregnancy?'[24] Duden distances her own project from such questions. The alternative way of evaluating the technology and its history that she has in mind would set aside the whole question of what a particular form of technology *does* and instead ask what it *says* about those who participate in its use and about wider society: 'what forms of thought, what modes of perception, what sorts of experiential orientation gain currency through its use and its very existence'. The forms of thought, the styles of perception and the existential orientations that Duden wants to discuss relate primarily to the maternal experience of pregnancy and our general picture of the child in the womb. For the moment, it will be enough to note that Duden, like Feyerabend in the passage we have cited above, speaks of changed forms of perception that go well beyond individual applications of the technology. Here, too, we have a case of the mere existence of a form of technology causing a generalized change to the cultural form of perception.

In what follows, we will examine both examples in more detail. However, before doing so, it seems important to say something about more established views of perception and technology, for it is against the background of certain conventional *idées reçues* that the novelty of the viewpoint on cultural history implicit in Feyerabend and Duden's work can be brought out most clearly.

The traditional point of view

Let us turn again to that conception of technology whose dominant strains are to be found in the anthropology of Arnold Gehlen and the philosophy of Ernst Kapp. Although a large part of contemporary research in the field of technology can be located well beyond the intellectual scheme Kapp and Gehlen helped to establish, there is no denying that that scheme still has considerable influence, not just on public conceptions of technology, but also within technological research itself.

Kapp, as we have already seen, belongs firmly to a tradition that reaches back as far as antiquity and its view of technology as an *imitation of nature* – a view that implies considerable (and justifiable) regard for what we might call nature's own technological achievements. Kapp sees the technician and engineer as students of nature: their aim is to free human beings from their dependence on nature by reproducing its achievements artificially, thus making the bounty of the natural world widely available and subject to human control. Nor is Kapp's theory without influence. One detects traces of something very like this sort of neoclassicism in the contemporary conception of the computer as a kind of electronic brain, and in several streams of research in bionics, for instance, in attempts to adapt the principles of plant growth to structural engineering, as well as in efforts to put the aerodynamic properties of animal body shapes to use in aeroplane design. Such examples demonstrate well enough that Kapp's general view is in no way simply false – it is certainly not unfruitful. However, it is also clearly unsatisfactory. One of the oldest of all forms of human technology, the wheel, ought to be enough to convince us that Kapp, at best, captures only half of the explanatory story. As Sybille Krämer has ably shown,[25] when we take artefacts like the wheel into account, technology looks much less like an imitation of nature, and much more like a material embodiment of social functions (in the case of the wheel, a response to the social demands for transport, traffic, communication, etc.).

The conception of technology as an imitation of nature has received considerable support from theories and examples relating to perception. Analogies between the production of images using lenses in optical devices and in the eye have played an important part in theories of technology from the outset, as has the similarity between the eye and the *camera obscura*. Even today, it is difficult to resist the temptation to think of photography as a kind of technical imitation of natural vision, or of the *realism* of photographs as somehow resting on an analogy between the camera and the eye as image-generating mechanisms. In speaking of the technification of perception, my suggestion is that the relationship between vision and photography is certainly not as clear cut as may appear; it seems necessary here to continue down the interpretative path suggested by Feyerabend and Duden and pose the same questions of photography that we posed of ultrasound and the telescope: how much have cameras *themselves* actually changed the cultural pattern of human vision? Moreover, could our sense of the realism of photography be a mere secondary effect of that cultural change? (This is to pass by an even bigger problem: whether photographs should themselves be regarded as analogous to visual images. When presented with a photograph, we can, after all, perform the same Gestalt switch that we can with the physical reality around us. Because the photograph too is a physical object, we can see it one way or another, whereas a visual image exists solely from a single point of view.)

On the imitation thesis, medical techniques of visualization, such as sonographic imaging, would have to be seen as a form of technologically enhanced vision, the thermometer as a technological enhancement of our natural sense of body temperature and the microphone as a kind of enhanced acoustic organ, while the production of (analysable, manipulable) data by each of the three devices would be seen as objectifying, sharpening and extending what already takes place in and through the process of sensory perception. The one-sided preconceptions of a philosophy of technology that tells us that technology is an imitation of nature touch at this point on another group of preconceptions we will discuss shortly. These are the one-sided assumptions of the *physiology of perception*.

According to the anthropological theory of technology, given its paradigmatic form by Gehlen,[26] technology exists first to reduce demands on the human body and second, to extend and substitute for bodily organs. The organ-extension thesis, in particular, draws on a rich stock of examples relating to perception. Both the telescope and the microscope are interpreted as devices for comprehensively broadening the resolution of human vision and thus extending the spectrum of what can be visibly differentiated; contemporary medical techniques of visualization, we note, could be interpreted along similar lines as techniques that extend the capacity of human perception. The thesis of 'organ alleviation' (viz. that technology exists to reduce the demands on bodily organs) is, to some extent, original to Gehlen;[27] however, in broad terms, it is traceable to Plato's *Protagoras*, where it takes its place within a wider conception of man as an essentially *deficient* being, as we have noted. In brief, Plato/Protagoras' idea is that man is particularly vulnerable to the influence of natural happenstance because of his lack of specific natural attributes and fine instincts, and because, physiologically speaking, he is born prematurely – three basic shortcomings which make it impossible for him to survive without technological assistance. However, for Gehlen, it is certainly not just tools or weapons, or even the general technical repertoire of material means, that make good the natural deficit. Social institutions in this story exist to fulfil the same function too.

On the issue of perception, the notion that man is a naturally deficient being suggests that he is exposed to an excess of indefinite sensory stimuli. Here it becomes particularly interesting to consider technologies of perception from the point of view of organ alleviation. The organ-alleviation thesis certainly has prima facie explanatory purchase. (Technologies of perception certainly make possible a sort of organ alleviation by establishing a technically manipulable distance between the object and the viewer. Photography, for instance, certainly seems to provide the contemporary tourist with the means of warding off excessive sensory stimulation in this sense.) However, Gehlen wants to take his theory further than that: the techniques that man, the deficient being, develops in order to reduce demands on bodily organs give rise to further techniques of organ *substitution*; a host of functions and capacities that man once had to take care of pretechnologically by brute physical means are now either delegated to technical devices or else become entirely obsolete. This phase of Gehlen's argument raises a range of questions of its own, questions that go beyond those we have already flagged, such as whether photography is an objective form of vision or whether sound recording using microphones is a technically mediated form of hearing. Here Gehlen confronts us with some more troubling dilemmas: whether photography simply reduces demands on human memory or is perhaps coming to function as a complete substitute for it, and likewise whether technically mediated communication simply reduces the strain of face-to-face communication, or whether it might tendentially be coming to replace it.

Those questions would undoubtedly be interesting to address in our own conceptual terms. For our present purposes, however, what is more important is Gehlen's conviction that neither the use nor the mere existence of technology has any effect on human bodily being itself – his implicit belief that man's bodily being is what it is *irrespective* of technology or technological development. This is precisely the aspect of Gehlen's work that limits the scope of his conception of technology, however practically relevant it may be in other ways. What we have already noted in relation to Kapp applies here too – the theory of perception that grounds Gehlen's anthropology gets no further than a problematic *physiology* of perception.

Labelling the physiology of perception as the *traditional* theory of perception may sound as dismissive as presenting Kapp and Gehlen's theories of technology as a source of our technological *idées reçues*. Yet we should certainly not underestimate the commonplace assumption that we *see straightforwardly what our eyes show us* or *hear straightforwardly what our ears take in*; nor can we ignore the boost that physiological theory has received from recent advances in the neurosciences. The theory models perceptual events according to a one-dimensional scheme of stimulus and response; since Müller and Helmholtz in the nineteenth century, physiologists of perception have posited the existence of specific currents of sensory energy, while more contemporary researchers point to the operation of internal sensory filtering and feedback mechanisms. Recent research has certainly succeeded in placing the facts of Gestalt psychology on a neurophysiological basis. Yet a number of prejudices remain: perception is still taken to be the exclusive work of mankind's natural faculties and hence something that can be known in natural scientific terms and no others; acts of perception are still rigorously compartmentalized as the acts of *isolated* individual organs. Perception is still held to be fundamentally passive; with the view of the perceptual process as divided between faculties comes the view that the function of those faculties is essentially *receptive*.

These, however, are philosophical presuppositions that deserve critical attention, for the question arises whether the very idea of the five human senses is itself a long-term product of the physiological approach which in its turn is being reinforced as acts of perception become technologically reproducible to an ever greater degree. Attention needs to be given to the formative influence of culture and, by implication, cultural history, on perception – something that is bound to elude research within the physiology of perception because the object of physiological investigation is bound to remain a culturally *preformed* organism. Finally, the simplistic notion of perception as pure receptivity is something that needs to be thoroughly rethought. Perception, as we will see next, is inescapably situational, and to that extent not only a receptive but a *communicative* deed.

In short, it is only when *cultures* of perception move to the centrepoint of theoretical research that it becomes possible to do justice to the part played by technology in structuring perception.

Cultures of perception

The concept of distinctive *cultures* of perception is what we will need in order to account for the role of technology in *actively structuring* perception. The notion presupposes that perception is not a mere physiological act and thus not anthropologically invariant: we must reckon with the possibility of historical changes to modes of perception and with the existence of different perceptual structures within different cultures. Moreover, if a particular kind of patterning of perception is an essential element of any culture, a particular kind of enculturation or socialization becomes essential if human beings are to *competently* perceive the world around them. It is meaningful to speak of the young being schooled to visually perceive the world in a particular way. Just as importantly, when one passes from one culture to another, a certain definite acculturation is necessary. In a foreign cultural environment, one is, for a time, *blind* to many things, and remains so as long as one is still learning to see. The physiognomic blindness towards Asian faces that afflicts Europeans (and its reverse among Asians) is well documented. However, our ways of seeing are not just relative to one or another formative cultural environment, but also to particular situations and practical contexts. Different forms of work habituate us to different ways of seeing and different practical contexts compel us to adopt different perceptual modes. A good example is the way modern forms of traffic strongly preference particular forms of perception in which *signals* rather than *things* give the visual landscape its primary contours. (And one can apply the word traffic in a very broad sense here, not just to street traffic, but to economic exchange or even interpersonal communication.)

Even such a brief overview of the more or less powerful cultural determinants of our ways of seeing brings into view an entire spectrum of alternative forms of perception: historical differences, cultural differences and situation-dependent differences account for some of the major variables of the way we look out into the world. But something else becomes clear as well. Over and above the sheer variability of human modes of perception, our overview suggests that the denizens of the world of contemporary modernity are each *individually* in possession of numerous different perceptual schemata and that they are hence able to switch between schemata to a certain degree. The fragmented multicultural nature of contemporary forms of life emancipates individuals from the constraints associated with dominant ways of seeing; indeed, one could say that our fragmented multicultural societies are what make the cultural preconditions of our ways of seeing detectable in the first place. Were it not for our freedom to manoeuvre between forms of perception, there can be no doubt we would be all the more tempted to take those forms as invariable natural givens.

The nexus between culture and perception has already been the object of extensive research by Rudolf zur Lippe and Dieter Hoffmann-Axthelm, whose books *Sensory Awareness* and *The Work of the Senses* explore their respective findings in some detail.[28] Disappointingly, the issue of the technological structuring of vision is not raised in either book, or, to put it the opposite way, neither zur Lippe nor Hoffmann-Axthelm reckon with technology as a culturally formative aspect of human existence. Nonetheless, we can draw on the compendium of approaches and concepts that zur Lippe and Hoffmann-Axthelm have put together in order to take the next step with our own inquiry. My focus in what follows will be on five concepts: articulation, Gestalt, perceptual 'keys', what it is to see or fail to see something and, lastly, the emotional colouration of perception.

The term *articulation* refers to the fact that in the absence of a practical framework, perception remains vague and ill defined; in short, it is repeated exposure to a particular practical context that organizes or 'articulates' the general perceptual field and thus brings it into definite focus. (Hence, incidentally, the close connection between linguistic and perceptual differentiations. The well-known example of the practical-cum-linguistic multiplicity of distinctions between types of snow among Eskimos is an obvious case in point.) Gestalt and articulation are correlate concepts. In opposition to the strictures of perceptual reductionism, since the 1950s Gestalt psychology has sought to show that perception has a self-organizing tendency that favours the formation of particular shapes. In the intervening years, neurophysiology has confirmed the Gestaltists' view and has, in fact, set it on a thoroughgoing neurophysiological footing. What is decisive for us, however, is that only a small proportion of the *Gestalten* that organize the visual field are present to the mind from birth. Far from being inherited, a Gestalt is a kind of learned pattern. Culture-dependent and shaped heavily by the individual's level of socialization, each Gestalt organizes the visual field differently. Hoffmann-Axthelm, however, has taken the notion of Gestalt-relative vision a step further with his concept of *social/perceptual keys*. In *Sensory Awareness*, he uses the term to stress the active moment of the perceptual process; perception on this picture is what almost literally unlocks individual situations – what gives situations a definite form in the first place. In actual fact, perceptual keys do double theoretical service in *Sensory Awareness* in also bringing to the fore the class-specific, gender-specific, age-specific and career-specific possible determinants of our ways of seeing.

Next, we have the subtle dynamics of seeing and failing to see. Following the lead of psychoanalysis and, in more recent decades, the studies of French philosophers such as Foucault, authors of various stripes have shown us that the process of perception is always simultaneously a process of *non-perception*, and that it inevitably entails an amount of disavowal and not quite wilful blindness. What is involved is something more than the lateral visual suppression connected to the process of perceptual articulation detected by neurophysiologists. Rather, perception and non-perception are bound up with the *affective aspect* of the way that we look at the world. Sight, but also other modes of bodily apprehension, is not only invariably directed by human interests, it is also shaped by fears and by emotionally or mythologically charged perceptual keys which close off the possibilities of perception in certain definite ways. (Hoffmann-Axthelm himself has been particularly keen to draw attention to phenomena of this sort in the field of sexual/corporeal perception.)

Instead of following these pointers any further, let us leave the general conceptual ground of our problem and return to the question we originally posed – to what degree is technology a factor in the cultural history of perception? What do the practical examples presented by Feyerabend and Duden have to tell us about cultural history and how might perception be thought of as culturally patterned in the light of their intriguing case studies?

The telescope: Vision and the theory of vision

Feyerabend's avowed aim in *Against Method* is to give those on the wrong side of scientific history their due – a goal that relates to his general view that there is no such thing as a *definitive* scientific method and that science has a much closer affinity to art than scientists and theorists have been willing to allow. The case study that is relevant for us is the introduction of the telescope as a scientific instrument, and in particular Galileo's attempt to present the newly discovered satellites of Jupiter to his scientific contemporaries by means of his new invention. As is well known, the latter proved either culturally or physically incapable of using the device, or they dismissed it as an improper means of demonstrating the existence of Jupiter's moons. Feyerabend's novel take on the episode involves rescuing Galileo's learned colleagues from the charge of brute ignorance and conservatism by showing that both of the parties to the debate had good grounds for their respective positions.

First, let us consider the claim that Galileo's opponents simply *didn't see* the phenomena placed before their eyes. Most of us surely appreciate that using novel devices, particularly optical devices, requires practice. For instance, hardly anybody who looks into a microscope for the first time sees anything definite at all. At least to begin with in any case, the problem is a lack of perceptual patterns. (This is why it is circular, though unavoidable, to paint a verbal picture of what young students using microscopes for the first time *ought to be* seeing, rather than encouraging them to discover what they have before them in their own terms.) Using a telescope may be comparatively straightforward – although perhaps the fact that telescopes are a more common part of the furniture of everyday life makes the matter slightly more difficult to judge. The problems we have with both microscopes and telescopes would seem to result, in part, from having to use one eye only and, in part, from the fact that when we look into them, it remains unclear for a time what distance the eye is to bring into focus – if the problem can actually be coherently expressed in those terms.

In his book *The Body of the Scientist*, Werner Kutschmann[29] has given us a striking picture of the historical *disciplining* of perception that needed to take place before scientific instruments could really come into their own as tools for obtaining meaningful scientific results. When we first come into contact with objects like microscopes and telescopes, each of us in a way has to retrace the path of this historical disciplining process. My thesis is that the introduction and diffusion of instruments like the telescope and the microscope leads, in the long run, to a certain prioritizing of sharp focused vision over all the various other potentialities of sight.

The second point that emerges from Feyerabend's account – viz. that (at least some of) Galileo's contemporaries refused to look at what Galileo purported to have discovered through the telescope, or else refused to accept the telescope as a valid means of determining scientific facts – has been the source of plentiful mockery down the ages. Yet at bottom all that this mockery goes to show is that between Galileo's era and our own a very basic change to our ways of seeing has taken place as scientific instrumentation has itself made its mark on human perception. Perhaps the best way to think of what is going on here is to recall that for Galileo's contemporaries it seemed open to doubt whether the telescope was a *bona fide* means of representing reality or an instrument of illusionism. From antiquity until well into the eighteenth century, the notion of mechanics at the heart of European man's understanding of mechanical instruments was quite different to our own; we have had occasion to note several times that, in keeping with the Greek meaning of the term mechanics, man-made machines were often thought of as tools of trickery. The cultural world of Galileo's day was rich in mechanical curiosities that were consciously designed to deceive, so couldn't the telescope have been an instrument designed intentionally to produce optical *phantasmata*? The problem was a particularly thorny one in the case of Galileo's new device because the object that it purported to bring within the range of scientific observation (a heavenly body) was, by definition, physically inaccessible to the observer. The next four centuries were to see a profound reversal of the relationship between the technical devices and the surrounding object-world – to the point where we are now accustomed to letting mechanical instruments tell us what astronomical bodies are and what they look like. At best, we might ask ourselves today if another type of instrument, viz. our own sensory 'apparatus', might also suffice to ascertain the facts of nature. To underline the point, we could add an example from a very different realm of sensory perception. Although the thermometer might have started out life as a device for quantifying sensations of relative hot and cold – this is precisely how the concept of temperature *first came into existence* – today we tend to talk as if our subjective sensations of warmth were simply a less reliable guide to determining a scientifically defined quantum.

But, to return to the telescope. As well as being unable to see what Galileo was showing them, some of his contemporaries were also uncertain whether the image that they beheld in the telescope might not have been produced by the telescope itself. Indeed, without an optical theory that explained the operations of the telescope – the theory itself was in fact supplied somewhat later by Kepler – Galileo had no real means to counter their doubts. This might, at first sight, appear to be a relatively trivial point – why would one expect a device to be suitable for establishing scientific facts until scientists are in possession of a theory that accounts for the workings of the device? Yet here things are not quite so simple, for Kepler's theory, designed specifically to explain Galileo's new mechanism, was itself to contribute to a wide-ranging new explanation of the phenomena of vision: in fact, Kepler's theory of the telescope was to become a *sui generis* theory of vision, displacing a host of previous theories. Plato's was one such theory; it held that vision was a two-way process, i.e. that visual images result from a confluence of the light of the eye and the light emitted by the objects themselves. A second theory that Kepler's contribution sent into retreat was Aristotle's idea that vision occurs within rather than by means of light. The new optics was thus to revolutionize European humanity's entire understanding of vision and, by implication, its entire self-understanding. In Kepler's work, vision becomes a relation of *reproduction* that takes place through the medium of light rays.

Yet Feyerabend is unwilling to sharpen his hypothesis that 'the practice of telescopic observation did not just change what one saw through the telescope, but also what one saw with the naked eye'; whatever radical change to our ways of seeing might have been precipitated by Galileo's invention of the telescope, Feyerabend himself hesitates to conclude that the visual data themselves were no longer the same. One wonders exactly why Feyerabend baulks at the latter conclusion and what a sharper form of his argument would entail. When he speaks of a change to what is seen with the naked eye, does that not mean precisely that what is seen before and after the change is something different? Similarly, Feyerabend has his doubts whether the Greeks and the Romans of the period between Anaxagoras and Lucretius really did perceive the definite outline of a face (i.e. a definite set of surface structures) in the moon. But when he turns to a later line of evidence, this time reaching from Aristotle to Plutarch, he convinces himself that the later ancient world did indeed discern the contours of a certain definite lunar physiognomy. From this point, two interpretative pathways seem to branch out: perhaps the perceptual change that takes place between earlier and later antiquity did not affect *what was seen* so much as whether one believed that what was seen was *important*; for instance, whether one considered it relevant to astronomy. Feyerabend's wavering between the two possible interpretations, on my reading, has to do with the very way that he posits this distinction between definite sets of visual appearances and their wider relevance. The way out of the impasse is surely to recognize that the articulation of the visually given is, in the most basic sense, a process of seeing *with a view to* something definite, i.e. a process of seeing the given as relevant in a definite context. Thus, if the context of relevance changes in the course of history, the articulation of the visual field will undergo a related change, as will the degree of definition with which phenomena are perceived. In response to Feyerabend, one must reaffirm that it does indeed make a positive difference whether human beings scrutinize the surface of the moon for the outlines of a face or the contours of lunar mountain ranges.

To sum up the changes to the perception of the heavens and to human vision as a whole that were brought about by the invention and increasingly widespread use of the telescope:

1　The existence of sophisticated optical devices like the telescope has brought about a situation in which *sharp focus* has become the chief yardstick of all visual perception.

2　The existence of sophisticated optical devices has annulled the privileges of unaided sensory perception as a means of accessing facts. Before early modern science reversed the relationship between man and his technological accessories, the achievements of mechanical artifice were measured against unaided human sense perception. Today, by contrast, sense perception has to measure up against the capacities of technology.

3　In today's world, the knowledge of reality as a whole that has been opened up by sophisticated optical devices is itself what provides the organizing structures and the criteria of relevance for unaided human perception.

Endoscopy, the child and the maternal body, bodily self-awareness

My second group of examples relates to new medical techniques of visualization that redefine the human body as a new kind of observable object. In her book *Women's Bodies as Public Places*, Barbara Duden has traced the broad outlines of the history of such techniques from their beginnings in the early modern anatomical studies of Leonardo and Vesalius through to contemporary applications in endoscopy. Duden's focus, however, goes well beyond the development of optical instruments designed to make the interior of the human body visible. *Women's Bodies as Public Places* takes in everything from anatomy to surgical technique, methods of dissection, chemical preparation and medical notation.

By *visualization*, here we mean something a little different from what was at issue in our discussion of the telescope, which revolved around changes to patterns of visual perception set in train by technological change. Instead of a shift taking place *within* a single sensory dimension, here we will be looking at a shift of emphasis *between* dimensions. The change that we want to focus on is the change to bodily self-awareness that results when the sense of touch and our associated intuitive sense of our bodily disposition are displaced by the sense of sight as the prime sources of our knowledge of our own bodies. The bulk of our attention will be on technological developments dating from the late twentieth century, for it is the latter that, in Duden's view, have fundamentally altered both general social perceptions of pregnancy and women's relationship to their unborn infants as they develop within the womb.

There are three new technologies that Duden sets out to study in detail: sonography (the visualization of the interior of the body by means of ultrasound); direct (internal body) photography and the use of optical probes in prenatal testing (a technique that Duden dubs '*in situ* foetus-copying'); and, third, the use of ultramicroscopes to visualize the fertilization of eggs and the early stages of cell development. For Duden, all three techniques are fundamentally techniques for visualizing what *in itself* is beyond the range of vision.

'Foetus-copying' is probably the most purely optical technique of the three – 'a case like that of the telescope, but more extreme', as Duden puts it,[30] yet still certainly a case of visualization in her sense, i.e. a technique for making visible what *in itself* is beyond the range of visibility. Duden's concern is that when one catches a (technologically assisted) glimpse into the body's interior, one risks losing all existential identification with what one is seeing: one sees oneself as something essentially foreign. Another way of putting it would be to say that what one catches sight of is no longer given in self-aware bodily sensation. (In Hermann Schmitz's terms, one could also speak of a switch from the perception of a subjective fact to the perception of an objective fact.[31]) Just as importantly, the images generated by internal body photography are much more of a technical *construct* than they immediately appear; in the literal sense, such pictures (at least those presented to the scientific laity) are a montage of numerous individual 'shots'. To quote Duden again: 'What actually comes into focus are small visually senseless segments of the interior of the womb. It is only when such close-ups are spliced together and often heavily retouched that the illusion first comes together of being able to see what remained hidden since time immemorial'.[32]

In the case of sonography and of images generated using ultramicroscopes, we are talking about visual constructs in an even more fundamental sense. The cellular activity that ultramicroscopes bring into view occurs in physical spaces far tinier than the wavelength of light. Duden is right to speak in this case of an act of photographic *creation*: 'This is not a process of visual representation that *makes use of light*, but a process of visual creation *out of light*. The object here is not photographed in light, but manufactured from light: it is a visual appearance that is in a certain sense made of light, a product of [technical] *photogeny*, not photo*graphy*'.[33] Duden's account makes it clear that we are dealing with visualization in the strictest sense of the word – a process of making visible what is, in itself, literally invisible. The same applies to sonographic images that are generated using sound waves rather than waves of the same physical type as light waves (electromagnetic waves). Effectively, an echo in the depths of the body is analysed by a computer and given visual form on a monitor.

Duden's view is that sonographic photographs are an essential factor in the profound changes that have been taking place in very recent times to women's very conception of themselves during pregnancy. In the face of intensive technification, what once developed inside and out of a woman's body as an expression of the life of the body (the child within the womb whose living movements the mother sensed) becomes an ever more precisely defined individual with a name and a variety of socially specified claims and rights. In her book, Duden reports her encounter with a pregnant woman, Joanne, which she takes to be characteristic:

A black-and-white photo the size of a post-card was passed from hand to hand between the five women. In the picture one could make out the schematic outline of a skittle-shaped object around whose mid-point two irregularly shaped shadows of different sizes could be seen. On the edge of the glossy card was the scale indicating the dimensions of the being whose picture the women ostensibly held in their hands. "That's Brendan. His growth is normal," Joanne said, comparing the ends of the two shadows with the key on the edge of the card. There she sat next to me with her swollen belly and explained to her visitors what they were looking at. The dark upper portion of the hazy object was a head, then came the child's stomach – the two shadows were probably Brendan's feet, which he was holding up against his stomach. "Oh, and look, look . . . here you can almost see a penis."[34]

The research into methods for visualizing the body's interior during pregnancy is, of course, shaped not just by the curiosity of parents, or the logic of technological innovation, but equally by contemporary debates surrounding abortion, prenatal diagnostics and genetic selection and by the requirements of scientific experimentation on zygotes and, in some cases, foetuses. What *Women's Bodies as Public Places* attempts to spell out is that changes to the way that we perceive conception, pregnancy and childbirth are inseparable from changes to perception in a broader sense; for Duden, they amount to nothing less than a profound shift in social consciousness and in our entire picture of the world. Embryos, foetuses and zygotes, in short, are not simply naturally given beings, but beings that are constituted through a particular technoscientific approach to conception, pregnancy and birth. Duden's work shows that the debates surrounding these issues – even when (or indeed *precisely when*) they are conducted in the name of protecting the life of the unborn – are fundamentally influenced by medical processes of visualization; the latter, she contends, have led us to think of pregnancy as something other than an expression of the life of the body, viz. as a *procedure* involving embryos and foetuses, which only makes sense in terms of a deeply abstract concept of human life. For the pregnant woman, the change is marked by a tendency to see herself as a kind of corporeal environment, a vessel that exists to support the development of a being for which she is responsible and which has rights that are quite different from her own, even before birth. In a work entitled *Women With Glass Wombs*, Eva Schindele has made the same point in sharper terms:

> Formerly the space that protected the identity of the unborn from any kind of intrusion from without, women's bodies have been reinvented as fraught, if not downright dangerous, while at the same time the living embryo is becoming a manifold object of scientific curiosity.[35]

In summary, techiques of visualization in the field of reproductive medicine are changing the entire experience of pregnancy, including the ways that women literally and figuratively perceive themselves and their unborn children. What is at issue here is primarily a transition to a new mode of perception, a shift to the primacy of the (technically mediated) sense of sight away from an intuitive sense of the inner workings of the body. The change brings with it a certain sort of distancing effect, a certain sort of objectification, and, indeed, a certain degree of alienation. Even in the many cases where the most advanced techniques are not put to use, technology is transforming *what pregnancy and prenatal life are* in the cultural imagination of the contemporary world. As a result, public discussion about the political and legal problems surrounding conception, pregnancy and birth no longer relates to bodily self-experience as such, but increasingly to the physical objects (fertilized eggs, blastozytes, foetuses, etc.) that are co-constituted by medical science and its technological apparatus.

Future research

Although in conclusion I would like to put the technification of perception in a broader perspective, it also seems important to mention some further examples, if only briefly, in order to give an idea of the breadth of the field of potential future research.

Thinking back to the examples that we have explored in some detail, it becomes apparent that our emphasis has been on *visual* perception. Of course, that is in part a reflection of the primacy of the sense of vision in our contemporary culture-world – something so entrenched that vision and perception are frequently taken as interchangeable terms. Swimming against the tide a little, it would be interesting to see what sorts of processes of technification might be taking hold in other fields of sensory experience.

We recall that our second case study relating to visualization and reproductive medicine made brief mention of our sense of touch and of our intuitive sense of the inner workings of our own bodies, while our first excursus dealt with the thermometer and its role in perceptions of heat (temperature, etc.). Far-reaching changes to our ways of *hearing* are also clearly underway in today's world; however here the relevant technologies operate in a quite different sense to visual technologies like the telescope and the microscope. In short, our ways of hearing are not being transformed by devices for extending the spectrum of audial experience, but by reproductive technologies that transform the meaning of sound, speech and music by multiplying the sheer volume of what we can hear.

Inventions such as the gramophone and radio were initially condemned by cultural critics, such as Adorno, as inadequate substitutes for the live experience of listening, though as we will see, Walter Benjamin's famous essay 'The Work of Art in the Age of Mechanical Reproduction' of 1936 struck a very different note, looking forward to a time when the new technologies of reproduction would give rise to forms of art and modes of perception of a radically dynamic new kind. *Pace* Adorno, modern technology has not, in fact, ended up reducing the demands on human hearing to the same uniform low level, but has raised them considerably. Moreover, the widespread use of audio technology has led to a situation where music has become all but ubiquitous; in some instances, for example in the case of the World Soundscape Project, the technology has also been used to draw imaginative attention to the omnipresence of sound as part of modern human life. Musically, advances in audio technology have led to the discovery of both 'raw sound' and the spatiality of music, to the aesthetic integration of everyday sounds and to an artistic prioritizing of sheer noise, though also to a rehabilitation of the myriad subtle qualities of the human voice. In fact, the technical-cum-cultural transformation of our listening habits is a subject that has received little attention from researchers and thinkers. As examples of what is possible in the field, it is worth mentioning research conducted in the 1950s into the transformation of music into a paradigmatic consumer good,[36] as well as Shuhei Hosokawa's essay 'The Walkman Effect'.[37] As the title suggests, Hosokawa focuses on the extensive use of the Walkman as part of modern urban forms of life, though obviously one could extend his analysis to iPods, MP3 players and any number of popular technological products that have come on to the market in recent years.

It would be possible to continue through the list of the five senses and point to similar developments in other domains of sensory experience. However, to do so would be to run the risk of simply reproducing the doctrine of the five senses, which, as we have pointed out, is both physiologically reductive and technologically overdetermined. In fact, the limitations of such physical theories of perception need to be rigorously examined as part of any self-critical attempt to further theorize the technification of perception.

Theorists interested in holding conventional physiological interpretations of sense experience up to the critical light need to insist that perception is something that takes place between the two experiential poles of atmospheric or synaesthetic sensory activity on the one hand and responsiveness to a world of signs and symbols on the other hand. Two of the starkest trends within the technoculture of the contemporary world are presented by the technological production of atmospheres and the predominance of technology within the world of image production, and yet the forms of human bodily experience that might develop under the influence of technically fabricated atmospheres and technically fabricated imagery remain entirely unclear. One of the possible alternatives can be summarized under the heading of *cyberspace*. Another very different alternative has been called *Neue Sinnlichkeit* – a self-aware rediscovery of the manifold possibilities of sensory experience. What is obvious, in any case, is that the entire perceptual life of human beings, in future, will be increasingly defined by the technologies of modern media and communication.

Let me conclude with an overview of the main ways that the technification of perception affects our world:

1 Modern science is defined, in part, by an attempt to distance its endeavours from sense perception, if not liberate itself from sense experience altogether. In the final analysis, what comes to count as a valid input into scientific

research is not the givens of sense perception, but what scientific instruments reveal under experimental conditions. In medicine, the trend is well underway. In large parts of natural science, it has already won out completely. For our analysis, what is important is that the transition to more and more instrumentally mediated data collection has far-reaching repercussions for human self-experience and daily life. To imagine the future course of developments, we would do well to recall the conceptual/experiential shifts brought about by the mainstream use of the thermometer or the present-day conceptual/experiential impact of biofeedback devices, like blood pressure monitors, which make up for users' lost trust in bodily self-perception by *telling them* how they feel.

2   The technification of perception has led to a devaluation of unaided sensory experience. Where once human sense perception supplied our basic terms for assessing the use and effectiveness of technological instruments, nowadays something like the opposite is true: instrumental access to reality has become the yardstick with which we define the capacities and limits of human sensory experience.

3   The technification of perception is bound up not just with a general trend towards objectification, but also with alienation and with a trivialization of experience. Technification does indeed lead to an increase in the intersubjectivity, reliability and precision of perception. Yet it also creates a distance from what is perceived, leading to a marked loss of affective participation.

4   The technification of perception is inseparable from the construction of new perceptual patterns and schemes. The perceptual keys that we use to organize the perceptual field either change or are created anew as new technologies of perception and new reproductive technologies spread rapidly throughout the world.

5   The technification of perception is also inseparable from the creation of perceptual worlds that are independent of material reality. If perception is, by its very nature, *mediated*, i.e. if what one perceives is perceived in and through one or another medium, then the invention and widespread use of technical media make it possible for perception to take place almost exclusively in technified fictional domains.

Sketching these sorts of trends towards the technification of perception implicitly means adding an element of science fiction to one's analysis – the above summary could certainly have gone further in this direction – for it is impossible to assess what impact the wild profusion of new technologies of perception and reproduction might have without extrapolating current tendencies along lines familiar to us from the work of novelists and film-makers. Until now, studies of technological development have mainly focused on technologies of economic production, transport and war. Yet it is precisely the technologies of perception and *re*production that will change what it means to be human in the decades and centuries to come.

Our picture of the effects of technological development on human sensory experience has already touched on several important ethical questions. Techniques of visualization in the field of reproductive medicine do not merely shape our visual picture of the world, they also profoundly affect our whole notion of conception, pregnancy, birth and human life, in doing so altering our sense of what morality demands and what it deems beyond the pale. The practical effects of these developments are particularly dramatic in the field of medical surgery. The widespread use of endoscopy in recent decades has, in a certain sense, made the fantasy of human transparency – the glass human being of popular political fear – into a realizable physical possibility. Surgery need no longer involve opening up the body. Working with an array of endoscopic probes, cameras and computer monitors, the modern surgeon moves about and operates on the interior of the body almost wholly indirectly.

Finally, we can point to the way that the technification of perception is changing the form of human life by changing the way that we go about the daily business of living. Considered in the broadest conceptual terms, sensory perception is, after all, a form of human *presence*, viz. bodily presence. But one of the prime effects of technification is to disrupt this sense of bodily presence, making it into something increasingly reflexive and abstract. The trend towards technological reflexivity was, in many ways, already evident in the integration of photography into daily life; the wild popularity of the snapshot was already beginning to change the basic forms of human experience in the mid-twentieth century. Although it was no doubt valued as a prompt to memory, the sort of cultural question mark posed by instant mass photography ought to be obvious: the snapshot presents something that can no longer be experienced, or at the very least is experienced differently, because of the act of taking the shot. Polaroid cameras and, more recently, mobile phone cameras, accelerated the trend, making the image itself into an even greater component of the situation in which it is generated; the experiential content that images were once thought to give us access to diminishes accordingly. Something similar applies to the use of handheld video cameras, which evacuate individual situations of experience, or rather displace it from the immediate situation to the evening viewing session in front of the television. In short, more and more of what happens is coming to be defined in and through its representation in an image. Hence the importance of being (literally and metaphorically) in the picture, and of making sure that events of every kind are captured in images – the by now completely routine staging of parties and biographical milestones specially for the camera.

The consequences of technification for ethics and for the forms of everyday life have brought us to a circle of problems that, strictly speaking, lies beyond the technification of perception. Yet because perception remains something deeply embedded in all forms of cultural practice and in the everyday routines of human life, moving beyond our initial problem seemed necessary to get a full picture of what the technification of perception might mean for us as human beings. Ethics has been our destination in the present essay. In our next study, it will be our point of departure.

Genetics, biotechnology and human self-understanding

'Ethics and Hypocrisy'

It was certainly an astute move to schedule a symposium entitled 'Ethics and Hypocrisy' as part of the 1998 exhibition *Genetic Worlds* held at the Federal German Exhibition Centre in Bonn. Could there be a more provocative way of thematizing a misunderstanding of the nature of ethics – or perhaps we should say a false hope placed in ethics – that is widespread among natural scientists? My point is not one about the naivety of natural scientists, most of whom came to acknowledge long ago that their expert knowledge might help in clarifying facts that are relevant to social problems but can rarely crystallize real-world decisions.[38] Rather, what I wish to point to is that this welcome modesty on the part of scientists brings another problem in its train: the problem that science now habitually looks to ethics for answers to the practical questions that it itself hesitates to answer. Ethicists on this view turn out to be nothing more than a different kind of scientific expert, viz. experts on values. And where the arguments of scientists or technical experts for or against particular practical decisions run the risk of hypocrisy, viz. when their arguments could be a smokescreen for the vested interests of third parties, it was ethicists whom the scientific fraternity typically turned to for pronouncements about moral right and wrong. Something crucial, however, got overlooked in the process: that where ethics develops into another specialist discipline – where the ethicist becomes a career professional like any other – ethics itself runs the risk of being instrumentalized: the arguments of ethical professionals become just as susceptible to the institutional or ideological pressures of those who finance their activity as the scientist who has delegated the problem of ethical consequences in the first place. Where ethics is shaped by external self-interest, it itself becomes hypocritical. Just how close we have come to such a position of entrenched hypocrisy can be gleaned from an interview in a 1998 edition of *Philosophy/Information*,[39] which openly accuses the bioethics fraternity (*not just* individual bioethicists) of systematically aiding and abetting biochemical big business.

Ethics and social expectation

The attitude to ethicists among natural scientists no doubt reflects a much more widespread social expectation – the fact that people are generally turning to ethics for the answers to otherwise unanswerable questions, which has, in turn, led to a quite extraordinary flowering of the subject, and to a kind of rejuvenation of philosophy too. Traditionally, medical ethics and social ethics have always played a part in the academic curriculum and, to an extent, in public discourse. Now, however, the number of recognized ethical specialities has multiplied to include bioethics, environmental ethics, peace studies and the ethics of technology. Ethicists are undoubtedly in real demand. But *why* exactly? Are our environmental problems likely to be solved by ethical means? Is the problem of world peace an ethical problem? Is the technological transformation of our societies and our lives something that ethics is going to help us to cope with? Aren't questions like these, in large part, questions of responsibility – a responsibility that the main players in the relevant spheres of social action must face up to? Don't the myriad social problems that we face call above all for political, technical or environmental solutions and a host of legal regulations? Is the continual talk of responsibility – for instance among European politicians – not a pretext for inaction, and does ethics in such a situation not become a sort of para-political alibi, a way of dodging problems that are actually quite material in nature? Is ethics really the panacea for all the problems that we confront, the all-purpose solution that it now sometimes presents itself as?

Confronted with such sceptical questions – and we have deliberately sharpened the sceptical tone – it is important to reiterate first that many of the problems that we face are indeed moral problems, or at the very least problems with a major moral component, and second, that ethics can indeed make a contribution to solving social problems. However, in suggesting that ethics has an important social role to play, we are initially not thinking of ethics in the sense of a professional skill practised by academic ethicists, but of ethics in the sense of an *ethos*, or, in plainer terms, in the sense of what is *customary*.[40] Custom (habit, etc.) – what philosophers since Hegel have called *substantive ethics* – is something that can have an influence on vast numbers of human beings, essentially because what is customary guides everyday action in an almost wholly inconspicuous way, without the need for explicit decision making. To be sure, custom can generate problems for societies. However, changes to custom can lead to solutions to social problems too – an obvious contemporary example is the way that our habits as consumers shape our society's relationship to nature and thus make a difference to the health of the environment. However, it would be wrong to think of professional ethicists – philosophers in the broad sense of the word – as having any particular responsibility or expertise on this score; knowledge of the socially customary is essentially the province of social psychologists, while passing on customs and the whole issue of *changing* customs is something that takes place incrementally in the home and in the classroom. Again, changing attitudes to the environment are a good example, for we certainly have environmental education to thank for the astonishing changes to standard patterns of behaviour that have taken place in recent decades in relation to the public use of the environment, recycling, waste, etc. (Although we should also not forget that a corresponding change to our use of resources, above all water and energy, is yet to take place.)

At this point, it is vital to note three things. The first is that what we are calling moral questions simply do not – cannot – arise within the domain of the customary; morality begins where custom no longer suffices or where custom has itself become problematic. (That indeed was the claim that we just made: that our social problems today take the form of genuine moral dilemmas.) The second point to note is that a moral question is a *question that demands a certain seriousness*. This I take to be definitional: things become serious for the individual, in a fundamental ethical sense, if he is confronted with a decision that will simultaneously decide what kind of human being he is. Likewise, things become serious for a society if it is faced with a decision that will decide what kind of society we live in – a decision that touches on how we as a society understand what it is to be a human being.[41] The third point is that moral questions that confront the individual can only be answered in the course of life, not through recourse to argument. Moreover, in order to answer them, the individual not only needs to develop a capacity for action, but also a certain capacity for *suffering*, what we might call an ability to let himself be *affected* by other people and the world at large.

This last proposition might seem rather surprising, given the centrality of action and the sorts of norms that govern action in almost all previous ethical systems. Yet the fact remains that having any sort of feel for morally relevant situations demands a certain openness or sensitivity, a certain passivity, perhaps we might say a *pathos* in the Greek sense of the word. In a civilization like our own, however, the tendency of most people is to shield themselves from the various kinds of impositions and impressions that they are presented with by the social world. People's willingness to let themselves be affected by the world around them diminishes accordingly. Under the conditions of an urbanized, technified, perceptually hyper-stimulated social life, fashioning any sort of moral existence thus requires the explicit education of moral sensibility.

However, the passive/pathic element of human life does not just return to the centrepoint of personal ethics in the guise of openness towards the *wider world*, for individual ethical life also involves a recognition of that part of *oneself* over which one has less than full conscious control; ethics here means integrating what has traditionally been called nature[42] into one's understanding of oneself as a human being. In short, there is simply no denying that this moment of what it means to be a human being will always exert a powerful influence over us. And yet with the rapid advance of medical technology, and the profusion of possibilities for medical intervention that such technology opens up, we find ourselves in a situation in which the nature we ourselves are, our bodies, have themselves become morally problematic.

The line that divides what we have to accept about ourselves as a physiological given and what can be manipulated at will is no longer as fixed as it once was and that situation has major repercussions for childbirth and the physical characteristics of children, for how one responds to one's own physical constitution, moods and dispositions and, indeed, for every aspect of life that involves suffering, illness and death. Nothing would seem to be beyond the range of intervention or manipulation, if not quite in the present, then in the not too distant future. What one is willing or unwilling to accept as an unalterable given of nature has become a matter for moral decision. Nor is the dilemma one that any of us can escape, though some may face it in more acute forms or in more dramatic circumstances than others. In order to preserve his dignity as a human being, everyone will, at some stage, have to set some sort of personal limit to the by now practically boundless possibilities for technical/ medical manipulation of his own body.

The moral questions that confront society as a whole have to be resolved with the help of new social conventions, and that includes everything from the creation of new customs and habits up to explicit legal regulation. This is where a discursive ethics has its place, for moral *argument*, though not constitutive of the moral domain as a whole, is precisely what is required for the establishment of new social rules, laws and norms. The decisive point, however, is that such argument *needs to relate directly to the social context in which it takes place*: moral discourse conducted without due attention to the relevant cultural setting, to the history of the relevant country or the broader civilizational situation in which culture and country find themselves, is, in our view, likely to be an idle academic exercise.

Moral arguments must thus be situated in relation to particular social traditions, shared experiences and existing institutions of authority (recognized bodies of law, etc.). By their nature, such arguments are thus never *totally* theoretically justifiable and only sometimes universal in scope. On this score, the type of discursive ethics associated with the name of Hans-Otto Apel is in the grip of a scientistic misapprehension; moral conclusions, according to Apel and his followers, are deducible from first principles. Our view is the opposite: moral arguments, for us, only make sense as *moral* arguments when they address a concrete political or social situation; they always have to speak to what potential parties to real moral discussion *already* hold to be right and wrong. To put the point in Aristotelian terms, moral arguments are not essentially scientific, but rhetorical or *dialectical* in nature: they take their bearings from major social *topoi* that function as the established bases of already existing moral agreement, as we will see next.

Moral questions raised by genetics and biotechnology

Because moral questions pose themselves precisely where custom and habit are not enough to provide regulative norms for action or where the regulative norms that they provide have themselves become problematic, technological developments give rise to a wide range of moral questions. While Günther Anders might be mistaken in thinking that human beings' moral capacities have come to lag behind their technical know-how,[43] it is nonetheless true that technological developments can rapidly make custom and habit redundant and hence give rise to moral questions. For individual human beings, what that might mean is that the amount of technological manipulation of their own bodies that they are willing to allow makes a decisive difference to the kinds of human being that they are. For society, what it certainly means is that the conventions that we agree on today to regulate the possible uses of biotechnology will make a decisive difference to our basic social understanding of what it is to be a human being with a certain human dignity and what it is to speak of human nature. What we stand in need of, fundamentally, is a reinterpretation of the notion of human rights and an extension of the basic social consensus (in Germany enshrined in the constitution) about the sort of society that we aspire to live in.

All in all, we are confronted with tensions that arise as new technological forms act on an existing cultural domain, and, in particular, the domain of ethical culture. In today's society, we can see the tension playing itself out particularly clearly in and around three particular problems that I now want to examine in a little more detail: (1) germline therapy and cloning, (2) the relationship between contemporary genetics and eugenics and (3) genetic mapping and the social uses of personal genetic data.

The use of *germline therapy* in cases of hereditary disease seems like an obvious step given that the latest treatments make it possible to treat a host of medical conditions at their very root. Yet public discussions of germline therapy, which in Europe led to an outright ban, saw the crystallization of a new moral *topos* whose import was that applying the new therapeutic techniques would seriously call into question a very basic element of our self-conception as human beings. The new *topos* has been given two names, *natality* (Hannah Arendt's *Geburtlichkeit*[44]) and *natural biological descent.* Germline therapy, used for however admirable a purpose, would run counter to both notions because it would run the risk of turning individuals, in principle at least, into a mere technical/medical product, thus rendering their individual conceptions of themselves incompatible with a sense of individual human dignity. As it crystallized in the course of debate, what became clear was that the *topos* has implications for other forms of reproductive medicine as well; in particular, it implies a ban on cloning. Cloning – it ought to be made clear from the start – is certainly a natural method of reproduction; a wide range of living creatures, plants in particular, are capable of reproducing themselves by non-sexual as well as sexual means. And yet the development of cloning as a practical biotechnological possibility led to the realization that (self-)reproduction of this sort runs contrary to our understanding of human dignity. Each human being – morally speaking – ought to be in a position to think of his life as a new beginning, and in a radical sense, in order to understand life in genuinely individual terms.[45] And indeed, he *is* just such a new beginning as long as he is conceived through the unpredictable combination of the genetic material of two parents. Understanding oneself as a new beginning becomes impossible, however, if one knows that one is a mere genetic subsidiary of another individual.

Considered solely in its own terms, there are several things that can be said in favour of eugenics – at any rate if one sees it in the broad context of the *project of modernity*. Since the time of Comenius' grand plans for the general education of humanity, the notion of the general betterment of human beings has been one of the most important dimensions of modernity itself.[46] In very broad historical terms, the ethical problem with eugenics arises because what was conceived in the seventeenth century as human improvement in the sense of *cultivation, moralization* and an *extension of the civilizing process,*[47] nowadays threatens to become an exercise in genetic engineering. The main argument that has been raised in Germany against eugenics in its contemporary biotechnological incarnation is the historical connection between eugenics and racism. However, that argument is by no means just an historical one. It is both historical *and* moral because it goes to the heart of German society's idea of itself: to set aside the Germans' historical experience of eugenics would mean, in essence, denying that the German Federal Republic owes its very existence to the defeat of National Socialism. This is not to say that the argument does not also have implications that go well beyond Germany. Within the United Nations, a new moral *topos* is becoming influential, most notably in the official elevation of the human genome to the status of a *common human heritage*. This *topos*, which harks back to ancient notions of lands and goods held in common, stands firmly in the way of eugenic interpretations of genetics insofar as (or precisely because) it can be assumed that every improvement to human genetic stock undertaken by biotechnological means would be undertaken in the interest of *particular* individuals or groups. In a culturally pluralist world made up of a wide variety of interests, opinions and ideals, a common consensus about how to improve human genetic stock is virtually inconceivable, and for that reason alone eugenics in its contemporary biotechnological form would in practice amount to biotechnological racism.

The possibility of genetic mapping – i.e. the compilation of information about the genetic constitution of individual human beings in the form of data – in many ways presents a similar threat to the conception of ourselves set down in international declarations of human rights. At a purely theoretical level, genetic mapping might have its merits. One positive implication, for instance, is that a long-standing ethical postulate, viz. that each and every human being is unique, has a firm basis in physical reality. However, our current position presents us with an equal and opposite danger, viz. that the individuality of individual human beings could be reduced to their genetic identity. (In criminology, the possibility has already become a practical reality in the form of genetic fingerprinting.) In short, the dignity of the individual human being has to be defended against the excesses of scientific naturalism. We have to firmly insist that the individuality of a human being consists of more than a factual physical constitution; a person, we might say, is in this sense nothing less than what he makes of himself – the indivisible human unity that he is capable of biographically endowing his life with. Giving the green light to genetic mapping for non-therapeutic purposes would undoubtedly lead to discrimination in any case. Here we might say that the threat to the dignity of man posed by contemporary biotechnology is analogous to that posed by physiognomy in the eighteenth century. The practical implications of Lavater's claim to be able to read the traits of individual character in an individual's facial features were thought to be worth thinking through, and worrying about, by his contemporaries and perhaps we can say that genetic mapping threatens to deprive the individual of his freedom or reduce him to set categories in a similar sense. The right response to it is to assert forcefully that every human being should be able to define his identity anew in terms that exceed the givens of genetic make-up. Yet the consciousness of individual freedom – the latitude each must be given to live in excess of categorial determinations – is not just threatened where social agencies are in possession of genetic profiles, but also where individuals themselves are in possession of certain sorts of information about their own genetic make-up. Knowledge of this sort could easily assume the guise of an inescapable fate. An individual who was burdened with such knowledge would be in a similar position to those heroes of Greek tragedy whose fates were revealed to them by an oracle – only as a rules he might well be something less than a hero.

As a final remark, we note that we have confined our discussion to moral arguments of a defined type. Everything that could be said about the costs and benefits of biotechnology or the associated risks and expectations, while it indeed has a place within public debate, is nonetheless beyond the scope of moral discourse about the topic. Any evaluative standpoint that limits its attention to costs and benefits, such as utilitarian ethics, can only be called an ethics in a misleading sense. Such a point of view is prone to confuse questions that can be decided by a shrewd weighing of claim and counterclaim with genuinely moral questions. The latter, we repeat, are of a different order from the former, for in answering moral questions we are deciding simultaneously, at the most basic existential level, what sort of human beings we individually are, and likewise what sort of society we are living in.

5

The technification of nature

Artificial nature

Introduction

The expression 'artificial nature' has an air of paradox about it. Does the phrase perhaps refer to some sort of transitional phase that we find ourselves in? To a hesitation we feel, or to an inconsistency? Are the wealth of examples of artificial nature all around us perhaps testaments to a nostalgic human desire to hold on to a world that is fast disappearing, a world of nature 'pure and simple'? Or are they signs of a reconciliation between art and nature, a future synthesis between the terms of an ancient dichotomy?

The natural and the artificial are mutually exclusive, or that, roughly, is how we have seen them: nature, until recently, was what existed of its own accord, art (the artificial) was the product of human invention or human labour. In the 1873 edition of Grimm's German dictionary, we read that 'the word *artificial* [*künstlich*] in essence now applies solely to art understood in contrast to nature'. 'Now' is the keyword here: Grimm's suggestion is clearly that 'artificial' was once also taken in a different sense, viz. as cognate with *artful* or *refined* [*kunstreich, raffiniert*], especially as applied to particular skills or knowledge. In this latter sense, one could once have spoken in German of the artful or artificial way that nature seemed to go about creating her works; certainly talk of the 'art of nature' was widespread before Grimm's day, especially in the eighteenth century. What that talk captured was the unity and purposiveness that men believed that they could discern in the natural world, the functionality with which nature appeared to be organized; in the eighteenth century, the idiom became a particular favourite in the field of 'physical theology' – in the copious volumes of natural history that took the unity, beauty and purposiveness of nature as signs of the existence of a wise and intelligent creator, a God who was also a divine engineer and who created the order of nature. The following passage from Sulzer's *Moral Reflections on Certain Types of Natural History* is a good example of the mode of thought we are talking about:

> Thus nature is the source of human invention. Are the inventions of human art – he said – anything other than imitations of nature? There is an old saying that our arts derive from nature – that it was the spider who taught men to weave and sew, the swallow and the beaver who taught him to build . . . . Yet these words of common wisdom still don't go far enough. For we can say that *all* the inventions of art are either derived from nature or are to be met with in nature in even more perfect form. Nature is the original workshop of all the arts, an inexhaustible armoury which men draw on to create artificial machines and which easily surpasses all that men have devised.[1]

Sulzer effectively interprets nature as the prototypical artist, or, more exactly, the prototypical artist-engineer. As an artist, man on this view does not proceed in any way differently from nature; in terms of achievement, his arts remain far inferior to hers. Nature, we might say, is for thinkers like Sulzer essentially a model. It is what human art and technology seek to imitate and what they ultimately derive from.

Imitating nature: Art, technology and artificial nature

We thus come to the famous theory of imitation, which holds that art and technology are imitations of nature and which was influential from antiquity until well into the nineteenth and twentieth centuries. How are we to understand this thesis? How are we to understand art and technology as standing in an imitative relation to nature?

A word first about origins. Apart from a handful of anticipations in Plato, the thesis goes back to Aristotle, who in his lectures on physics puts the point as follows: 'In part technology [*techne*] perfects what nature has failed to complete, in part it imitates nature'.[2] The Greek term translated here as imitation is *mimesis* – a concept that has undergone something of a Renaissance in twentieth-century aesthetics, particularly as a result of its reconceptualization by Adorno. It is worth bearing one thing in mind here though: our twentieth-century notion of *mimesis* picks up on one particular sense of the Greek notion that played a relatively small role in the traditional doctrine of imitation.

A mime is, of course, an actor who enacts a particular type of representation, possibly something merely allegorical or whimsical. However, apart from the art of the mime, *mimesis* can also mean something like reproduction, copy or imitation – doing as someone else does in a quite general sense.[3] With this in mind, we can restate Aristotle's version of the imitation thesis as follows. Art and technology reproduce natural effects; in art and in technological artifice, human beings are doing as nature does. Relatedly, Aristotle thinks, as did Plato before him, that the visual and literary arts, unlike the art of music, *re*present what nature presents in its original form.[4] Technology, on the other hand, copies nature in quite a different way: the products of the craftman's workshop do as nature does in the sense that they are *teleological*. Aristotle makes the point by saying that if nature were to set about constructing a table, she would proceed as the carpenter does, seeking out appropriate material and putting it together as the desired aim, or *telos*, requires. Aristotle's concept of imitation is rooted in his teleological interpretation of nature. Of course, nature itself knows no tables, so here, as elsewhere, the products of technical artifice are to be understood in contrast to the works of nature. Aristotle's idea is that they *outwit* nature, forcing her to produce movements and effects that go beyond her natural tendencies, or are opposed to those tendencies: what the Greek designates as *para physis*.

Aristotle even wants to establish ontological grounds for the opposition between nature and technology that opens out at this point: nature and technology, in this respect, represent two quite different types of being. A being whose essence is natural contains its own principle of movement – it is responsible for its own development and reproduction – while a being whose essence is technological receives its principle of movement from human beings: it originates in and is organized according to a system of human purposes, and develops according to the plans and wishes of its human creators.[5]

So it would seem that in its original formulation the imitation thesis in no way claims that technology copies nature and her works. According to Aristotle, technology merely *proceeds along similar lines to nature*, viz. teleologically. However, the advent of scientific modernity changes all that, slowly but radically. The more nature comes to be understood *mechanistically* in modern scientific terms and the more forceful the scientific repudiation of nature's *overall* purposiveness becomes, the more science is struck by the extraordinary aspect of nature and her works, the more alive it becomes to nature's internal organization and the apparent purpose of *individual* natural processes. The imitation thesis is now subject to radical reinterpretation; the *ne plus ultra* of technical perfection as understood by scientists, craftsmen and engineers becomes the mechanical reproduction of nature's works. We have already made mention of the craze for automata that gripped the eighteenth century – a cultural moment crowned by the creation of a mechanized duck[6] – and today the goal of imitating nature as it was pursued by the automatic arts of the eighteenth century seems absurd. On the other hand, the solutions that nature has found to particular problems still seem exemplary to us today, and so we continue to study nature from a purely technical point of view – not just for the sake of pure knowledge, but also in order to make good heuristic use of nature's quasi-technological artifice for our own technological purposes. (The scientific field that devotes itself specially to the task is bionics, also known interestingly in English as biomimicry or biomimetics.)

Yet the imitation thesis also has a resonance well beyond these essentially pragmatic endeavours. Technology has been interpreted *theoretically* in the light of the thesis until quite recently as well, for instance in Ernst Kapp's *Outline of a Philosophy of Technology*.[7] Admittedly, the imitation thesis appears in a highly unusual form here: on Kapp's picture, technology is essentially anthropomorphic, a projection of human bodily organs. Nor does such a projection begin with a precise study of the organs of the body and then proceed to create technological copies. Technological imitation of nature for Kapp takes place rather at an *unconscious* level. Human action generally, and the activity of engineers in particular, involves an unconscious externalization of our inner being, which, once accomplished, allows us to recognize ourselves in the external world. Kapp's reading of late nineteenth-century technological innovations is typical here: 'Just as the nervous system is a network of cables running through an animal body, so the telegraph cables our civilization is today laying out in all directions are the nerves of humanity! How could they not be when the characteristic feature of technological organ-projection is the unconscious way in which it proceeds'.

To be sure, this is no straightforward reprise of Aristotle's interpretation of *mimesis*. The idiosyncrasies of Kapp's view also mark him out as an heir to the romantic philosophy of nature. The latter we can characterize in terms of three basic beliefs: first, that spirit manifests itself in matter, second, that nature is active unconsciously in creative human beings and, lastly, that in the works of human creativity, and in art especially, spirit becomes conscious of itself. Technology, too, can be interpreted on this picture as another kind of unconscious manifestation of the organic structure of human life. Kapp himself even thought that interpretation was confirmed by the biology of his day (e.g. by the discovery that the lamella structure of the human thigh bone obeys something very like the principle of bridge construction in its distribution of loads and tensions). Of course, it takes no special insight to see that there are more sober-minded ways of describing the relationship between the technological and the organic world: to satisfy particular functional requirements, technology has a way of using the laws of nature to create optimally effective forms similar to be those that are to be found in organic nature.

Kapp's theoretical edifice has a certain rococo imaginative appeal, but as a theory it is clearly confused. However, it was by no means the last of the modern versions of the imitation thesis. In his 1957 work, *The Soul in the Technological Age*,[8] Arnold Gehlen set about giving *mimesis* an anthropological sense of his own, this time defining technology not as organ projection but as organ substitution. For Gehlen, the imitation principle accounts for technology's role both in reducing the demands on human bodily organs (something we can call 'organ alleviation'), and in strengthening our bodily organs' basic capabilities. Technological development is thus a continuation of the biological development of the body, and, in its later phases, progresses from organ substitution to a totalizing substitution of the entire organic world.

Imitation in this context does not mean *direct* imitation, but functional imitation, if need be with thoroughly non-organic means. What Gehlen overlooks though is that technology is well and truly capable of fulfilling *non-organic* functions like social functions. Technified processes of production, technified methods of transport and technified media of communication are difficult to subsume under Gehlen's concept of organic function; moreover, they are precisely the technological forms that might seem most typical of modern technological development. In Gehlen's *chef d'oeuvre*, one detects a similar urge to vaguely extend the scope of the imitation theory that was already noticeable in Kapp: 'Our account of organ-projection', Kapp had written in his *Outline*, 'moves from artefacts which flagrantly imitate nature's tools to machines whose far-reaching basis in theoretical abstraction hides from view all blatant use of natural materials and natural means'. For our purposes though, the critical weaknesses of both theories are less important than their basic line of interpretation: the way that each reads technology through the idea of imitating nature.

How do Kapp and Gehlen's conceptions relate to the issue of artificial nature? In short, they each invite the conclusion that there is, in fact, nothing paradoxical about the notion of artificial nature in the first place, because they each hold that imitation of nature is actually at the heart of the technological enterprise. Works of technological artifice belong, by definition, under the rubric of artificial nature. They exist to recreate, using human means, what nature otherwise accomplishes on her own. Strengthening or substituting for the works of nature, or hitting upon functional equivalents, is the very point of the technological enterprise.

Moving from technology to art, we again find the imitation thesis playing a pivotal role, although the aesthetic version of the thesis has a rather more eventful history behind it. The notion that art has an obligation to imitate nature was given its classic modern formulation in Batteux's eighteenth-century aesthetics,[9] which set out quite deliberately to challenge the baroque aesthetics of elaborate artificiality that were characteristic of the era of Rubens and Bernini. Batteux's classicistic strictures held good until well into the nineteenth century.

As in the realm of technology, the aesthetic principle of imitating nature could mean either proceeding along similar lines to nature or reproducing natural effects.

Detailed exposition would take us too far afield. However, we can sum up the broad course of historical developments in the words of Kant's *Critique of Judgment*: 'art can only be termed beautiful, where we are conscious of its being art, while yet it has the appearance of nature'. What Kant is suggesting here is that art produces objects and forms that ought to appear to be natural: art exists precisely in order to create artificial nature. For Kant and for the neoclassical aesthetes of the eighteenth century, as much as for technologists like Kapp and Gehlen, artificial nature is in no sense something disconcerting or disturbing. Rather, it is precisely what art ought to strive for.

Neither technology nor art are artificial . . .

This brings us to a different tradition of interpreting technology and art, which tells us that the products of technological and aesthetic artifice can *not* be meaningfully thought of as artificial, let alone as instances of 'artificial nature'. Let us turn first to technology.

The history of modern technology has its origins at the time of the invention of modern science, the period covered roughly by the life of Galileo. The age of modern technology begins with the *abolition* of the opposition between nature and technology. Galileo inaugurates the decisive change of evaluative standpoint. At its heart is the view that the natural scientist's experimental apparatus, and by extension technology generally, does *not* force nature to perform movements or exhibit phenomena that are contrary to her natural tendencies. What takes place when we make use of technological devices is in no way contrary to nature. Quite the opposite, it gives us a far cleaner, purer picture of nature than we would otherwise have got by observing external nature with the unaided eye of everyday experience.[10]

Admittedly, Galileo's radical new point of view implies a changed view of what constitutes a natural tendency. *Pace* Aristotle, since Galileo, physical bodies are no longer regarded as having a natural tendency to move towards the centre of the earth; rather, they have a natural propensity for inertia – a natural tendency, we might say, to move according to the laws of inertia and gravity.[11] The decisive point for us though is that the new point of view depends on a new understanding of the notion of mechanics, and thus of what a mechanism is. For the Greeks, mechanics was a field that stood *outside* science. On the new scientific paradigm, mechanics effectively becomes the first and most essential part of science. And so it will remain until well into the nineteenth century.

The unity of natural science and technology that is characteristic of modern science from the very beginning is implicit in the elevation of mechanics to the status of *the* prototypical natural science. Modern natural science defines itself as the investigation of nature under technical, i.e. experimental, conditions. Its subject matter is no longer the world of external nature, nature as it is factually given, but anything and everything that is possible in accordance with the laws of nature. As a result, external nature itself comes to be seen as a mere partial instantiation of nature as a whole, while nature in the sense of the *naturally possible* is constituted as a substantially broader field. Technology, on the other hand, becomes one and the same thing as nature; rather than imitating nature's works, it is now seen as realizing what the laws of nature make possible no less fully than external nature itself. The example that immediately springs to mind is the wheel, though perhaps the car or the aeroplane make the point even more vividly.[12] Neither wheels nor cars nor planes have any sort of model in the natural world, unless it is claimed fancifully that cars imitate the motion of animals or planes the flight of birds. The only meaningful sense in which one could speak of imitation in this context would be purely Platonic, for on Plato's view nature as we access it through the senses is itself already a copy: a *r*epresentation of the eternal unchanging world of Ideas. Yet even in Platonic terms, it would be mistaken to think of technological artefacts as *artificial nature*. In Plato, technological products too are nothing more than copies; things that craftsmen produce, such as tables and chairs, are as much representations of eternal ideas as natural objects like horses and trees.

So it would seem that the notion that technology is an imitation of nature is by no means obvious or self-evident, and, in particular, that it is quite untrue to modernity's core scientific understanding of the nature of technology. However, the imitation thesis has also been undermined in non-scientific quarters ever since the dawn of the modern age. Hans Blumenberg[13] points us to a fine passage in Nicolas of Cusa's *De Mente* in which a figure known as the *idiota* (the reference is to a private citizen who practises a craft, not a fool of any sort) takes up culinary arms against the imitation thesis: 'Apart from the idea we have of it in our minds, a cooking spoon can have no archetype'. A little further on, the same personage continues: 'In fashioning the spoon, I refuse to acknowledge that I am imitating the figure of any sort of natural object'.[14]

So much for the distance that modern science and technology have placed between themselves and the imitation thesis. However, one could also point to a parallel development in modern art since the nineteenth century. Again, one example will have to stand for a detailed exposition – this time a short passage from the writings of the painter Paul Klee. Klee's reflection on the nature of his work brings out the analogy between modern science and modern art's standpoint on nature very lucidly, for his idea is that for art 'the visible world is nothing more than an isolated example, a small portion of the world in its totality; numberless truths of a completely different kind lie dormant in the world as it is given to us'.[15] For Klee, art cannot be an imitation of nature – certainly not an imitation of the factually given visual world, because 'art *makes things* visible rather than presenting us with what is already visible'.[16]

What is 'artificial nature'?

Our argument thus far would seem to suggest that on a conservative reading of technology every product of technological artifice constitutes artificial nature, while on a progressive reading it is largely meaningless to describe technology in such terms. At a pinch, we could maybe combine the two competing interpretations and say that when technology, consciously or unconsciously, imitates nature, then it *does* indeed constitute artificial nature, while when it brings latent possibilities to actuality, as in modern experimental science, it bears no essential relationship to nature as it is factually given to us.

Next we have to consider a different type of artificial nature, viz. products of artistic or technological artifice that are labelled artificial nature for quite obvious reasons in everyday speech, often with an undertone of irony or criticism. I will start by drawing up a list of the products that I have in mind and then try to clarify their ambivalent position in the contemporary cultural imagination.

First, we have the use of nature for aesthetic purposes. The aesthetic environment of our contemporary world, we note, is populated by artificial flowers and trees – artificial plants of all kinds, all of which reappear in ornamental form in architecture and design. Children's bedrooms and playrooms are brimming with stuffed toys, most of them animals. The nineteenth century, we recall, created artificial paradises for itself in the form of winter gardens and oversized glasshouses[17]; modern-day shopping centres, leisure centres, resorts, not to mention casinos on the North American model, create similar effects, often on an even more grandiose scale.

Artificial landscapes fit into a second, though closely related, category; we note that nowadays we often construct them over the top of mining waste heaps and rubbish dumps, sometimes in the middle of deserts. Environmental reclamation should also be mentioned here. It ranges in scale from local revegetation programmes to the restoration of stream systems reduced to the status of drains by urban or agricultural development, all the way up to continent-wide land-care schemes such as have been mooted in southern Australia and Amazonian South America. Beyond the horizon of immediate feasibility, one could include a brace of curious schemes that, for the time being, belong in the realm of science fiction: cities capped by glass domes in the earth's polar regions and notional human settlements on distant planets.

A third major subcategory of artificial nature embraces the technological regulation, reproduction and remodelling of the human body, the spectrum of innovations that stretches from artificially induced sleep to artificial breathing devices and dialysis machines, as well as the whole field of artificial organs, or prosthetics, starting with false teeth and ranging all the way to artificial limbs and kidneys, perhaps one day even artificial hearts and complete suits of artificial skin. The burgeoning field of reproductive medicine clearly belongs here too. Nowadays, *in vitro* fertilization is only one of medicine's most basic methods, and indeed the day may not be far off when it will achieve the goal of creating artificial environments with all the technically requisite properties of the maternal womb. Lastly, we ought to mention artificial intelligence, as little as some of its advocates would like to see it on a list of schemes for manipulating the givens of the human body.[18]

For a fourth category of products that meet the common definition of artificial nature, we can take synthetic materials (fibres, building materials, fuels, etc.) as prototypical. Synthetic elements (Americium, Nobelium, etc.) probably fit under this fourth heading too, Artificial life forms (Ventner's *Mycoplasma laboratorium*, etc.) certainly do.

Why then do we speak of such things as artificial? What sets these four groups of technological products apart from others? To be sure, no one would be tempted to distinguish between artificial and non-artificial cars or spoons or wheels (and synthetic elements, which we just allocated to our fourth subcategory, are not normally called *artificial* elements, but *transuranic* elements). Perhaps there is no great puzzle here though. In contrast to more straightforward technological objects like cars, spoons and wheels, which realize possibilities allowed for by the laws of nature, the types of things that one gives the name artificial nature to are those whose look, function or design clearly take their bearings from the factual givens of external nature.

A second problem poses a little more difficulty: why do we take our bearings from the factual givens of nature in the first place? The question obviously becomes more acute when we recall that modern technology no longer sees itself as bound to external nature as a model in any essential sense. Are our four subsets simply to be thought of as four different species of cultural anachronism? Should we think of them as together representing a mode of technological production that is yet to cut free of traditional associations and affinities, something comparable maybe to the various throwbacks in automotive history to the first generation of cars that still had some of the visual hallmarks of horse-driven carriages[19]? Maybe our problem has different answers for each of the subcategories of artificial nature. Starting with the *aesthetic* manifestations of artificial nature in our first subcategory, a detailed analysis would seem to show that the inhabitants of developed civilizations have a pronounced need for nature, which actually *increases* with increasing isolation from the natural environment, but which can be satisfied in the form of aesthetic substitutes: hence the abiding cultural significance of what I have elsewhere called the *gesture of naturalness*.[20]

In the case of our second subcategory, artificial landscapes, land-care programmes, etc., we need to explain the way that we take our bearings from nature a little differently. Here, the decisive fact we confront is that, despite the profound influence that human beings have had in shaping the world of external nature, the terrestrial environment on the whole remains something natural in a significant sense. Any human creation that is to endure for any length of time has to somehow accommodate itself to planetary nature as a whole; nature, we know, inevitably catches up in the long run with anything that fails to do so. This is perhaps why our efforts to shape the landscape around us are so often devoted to the restoration of natural environments. The experiences that we have tallied up in our efforts to master nature are crucial here, for what has become clear in the long run is that human mastery of nature does not free human beings from labour in any simple sense, but burdens them with extra labours; direct and indirect damage to the environment calls for the use of yet more technology to facilitate repair and further control, sometimes just to stabilize the environments that we have so thoughtlessly instrumentalized to begin with. In the long term, returning the environment to something approximating its original natural state often turns out to be economically less costly than further rounds of technological management and control.

Our third group of examples of artificial nature comes under the heading of prosthetics in the broad sense. Here, too, a plausible analysis suggests that we take our bearings from the factual givens of nature. As embodied beings, human beings are themselves *always already* natural beings, and prosthetic manipulation of the human body orients itself around this aspect of human naturalness for obvious reasons. The danger, given the pace at which technology is advancing, is that we lose sight of the horizon of our own natural being: in contemporary biotechnology and reproductive technology, and in efforts to create various hybrids of man and machine, human bodily nature is clearly ceasing to be a point of orientation.

Our last group of examples is in some ways exceptional. As in the case of artificial elements, we could simply say of synthetic materials or artificially created life forms that they actualize possibilities that are theoretically already implicit in nature – that, in short, it is a misnomer to call synthetic materials *artificial*. If, however, we continue to speak of such things as artificial, that is obviously because our basic belief here, or one of our basic beliefs, was that certain materials and certain species exist as it were with nature's blessing. What was naturally given could be divided, for instance according to Aristotle, into organs, species and homogeneous materials (elements): taken together, the three constituted a kind of fixed stock of living and non-living beings that remained unamenable to technological manipulation. Nowadays, when human beings create new materials or new species, they *artificially* extend the fixed natural stock of old. In doing so though, they take the givens of nature factually as a point of orientation, as they must if their attempts to extend the suite of already actualized possibilities are to take place within the framework of the existing natural stock and its established natural subdivisions.

In summary, our overview would seem to suggest that it is not technological products per se that merit the title artificial nature, but only those technological products that take nature in its existing state as a yardstick. The fact that nature functions as a technological norm undoubtedly explains the ambivalent status of such technological products. And, as we will see next, that ambivalence gives rise to widely varying views of the value of contemporary technology, including numerous criticisms.

Against artificiality

In general, we can say that critical responses to artificial nature are aesthetic, political or ethical.

In themselves, aesthetic responses are independent of social position and fashion. In his *Critique of Judgment*, Kant notes, for instance, that human beings have a basic interest in nature and naturalness, while at the same time suggesting that that interest presumes a certain level of education.[21] Today, as in Kant's day, the aesthetically educated strata of society would probably find artificial flowers in the home positively distasteful. Yet in courtly circles during the European Middle Ages, artificiality was exactly what was prized in questions of art and taste. European Mannerism is another case in point; art history has seen all sorts of revaluations of the value of the natural and the artificial. In the contemporary world, both tendencies are equally present. Some artists have a preference for natural materials and do their best to display them in all their subtlety and power. Others take enthusiastically to plastic, fibreglass and acrylic paints.

Yet there are other areas of life, well to one side of the art world, where the question of the natural versus the artificial presents itself as a question of *taste*. There was a time not so long ago in many Western societies when it was seen as uncouth to breastfeed babies, when bottle feeding and hospital birth were regarded as chic and progressive. As is typically the case, these kinds of value judgements followed the gradient of economic position and social status: the positive valuation of artificiality first took shape among socially privileged groups and was then embraced lower down the social scale.[22] Today we find ourselves in the midst of a countermovement. The period since the 1980s has seen a definite turn to less artificial lifestyles something evident in everything from nutrition to clothing to home birth. The tendency, moreover, seems unrelated to class or social status.

Turning to politics, we find that the positive preferencing of the natural over the artificial is based generally on a view of technology as a form of *domination*. One notable example here is the feminist resistance to conventional reproductive technology. Today we find many women contesting the technification of natural reproduction not under the banner of nature, but because they see reproductive technology as an extension of *male* control over the *female* body.[23] Yet this is not the only area in which resistance to artificial nature invokes values other than 'the natural'. Organized resistance to artificial nature tends to look beyond 'natural values' towards broader sets of political values because the domination of nature is inseparable from the domination of human beings over other human beings[24] and because the hazards involved in humanity's technological mastery of nature are distributed unevenly within society and across the globe.[25] Because the project of technological mastery of nature is a *social* project, acknowledging that it has become problematic, or indeed redundant, calls for a society-wide revision of basic attitudes. In that sense, it is not in the least surprising that the conceptual opposition between the natural and the artificial has become a source of political controversy.

However, both the explicit invocation of nature and the wave of open resistance to excesses of artificiality have their main form of expression in ethics, and, in particular, in a kind of morally coloured theology whose viability depends on underwriting the idea of a *given* order of nature – be it a given level of biodiversity or a given definition of human nature – with a religious notion of a *created order of being*. (The intellectual underpinnings of the Vatican's ongoing opposition to reproductive technology of almost all kinds are precisely of this sort.) What though, we might ask, is the moral opposition to artificial nature to base itself on in the absence of a religious guarantee? The difficulty of the question should certainly not discourage those looking to respond to the problem of artificial nature in secular terms; the answer would seem to lie in some form of human self-conception that is both open to public debate and amenable to a reasonable degree of social consensus. The computer scientist Joseph Weizenbaum, for instance, poses the question of desirable limits to artificial intelligence in this spirit. For Weizenbaum, what ought to be of prime interest in social debates about artificial intelligence is *not* what human intellectual functions can be taken over by machines in point of fact, but what human intellectual functions ought to remain the preserve of actual human beings in point of ethical principle.[26] The answer to the latter conundrum is, for Weizenbaum, not far to seek: human beings have a responsibility, viz. to themselves and to human dignity, *not* to divest themselves of certain types of judgement and decision making. Importantly, within such an interpretative framework, what human nature is – what ought to be maintained as an inviolable given – is to be maintained as such *by fiat*. There is no denying that here moral resistance to encroaching artificiality takes a paradoxical turn: a prime sign of human *self-determination* becomes a capacity to accept various facets of ourselves and of external nature as *unalterable natural givens*.

Such an alternative path of resistance to artificiality would seem all the more viable if it were to be embedded in a new notion of what it means to be human which pays particular attention to our own bodily nature. Having a body (or, as we have put it, *being* a body) after all means living in accordance with nature as an essential facet of one's own being. And that, in turn, means paying due heed to something foundational in oneself that cannot be technically produced, reproduced or manipulated.

Goodbye to the human state of nature

We have, perhaps, finally touched on the crucial point that makes the question of artificial nature so controversial, both aesthetically and politically and, above all, ethically. Ultimately, what is at issue here is human beings' core self-understanding, or, in other words, the very basis of the way that we interpret ourselves both in relation to external nature and the nature we ourselves are. To clarify the meaning of the question a little further, let us conclude with a thought experiment. What does the future of humanity look like if we picture the history of artificial nature, i.e. the technological history of nature, advancing at something like the same pace as it has in the past? The problem requires us to be flexible with historical timescales.

The earth, as we know, is several billion years old, and life on earth has existed for roughly one or two billion years. Mammals first appear in the fossil record of the Tertiary Era some 70 million years ago, while human beings make their appearance in the Holocene roughly a million years ago. Compare that with the history of technology: its first main periodic division begins with the Neolithic Revolution somewhere between 20,000 and 10,000 years ago. Subsequent millennia pass relatively uneventfully, but as we approach the present day, the tempo of historical change picks up ever more dramatically. Writing and the wheel are approximately 3,000 years old, the steam engine and machine tools are a little more than 200 years old. Telecommunications, artificial fertilizers and refrigeration are datable to the mid-nineteenth century. Computers, radar, lasers and biotechnology are all products of very recent history: all are inventions of the later twentieth century. Now, if one sets the multiple billion years of geological history alongside the 10,000 or so years of technological history – the mere millennia that make up the history of artificial nature – and if one bears in mind the scale of change to the planetary environment that those millennia have brought with them – then a dizzying prospect opens out: imagine the history of technological development extrapolated on geological timescales. In our current age of breakneck progress, we have got used to technological prognoses looking forward as far as the year 2020. Anything extending the notional horizon out as far as 2030 has a slight air of futurological crystal ball gazing. But simply for the sake of world-historical perspective – what sorts of developments might be possible if we extend the future of technology out by a further, say, 1,000, 10,000 or even million years?

Taking such a long view, even if in imagination only, ought to make clear just how much is at stake in the question of artificial nature. On geological timescales, almost anything might become technologically possible, from the settlement of distant galaxies to a division of the human species into artificially created subspecies adapted to different conditions. Life forms that are only able to reproduce themselves in constant symbiosis with machines are certainly conceivable, and so indeed is the complete dissolution of the human species as we know it today. On such a long view, the notion of artificial nature comes indeed to signify an intermediary phenomenon, a limit, perhaps the juncture at which an evolutionary decision has to be made: the decision whether humankind is ready to bid goodbye to the state of nature once and for all.

The expression *state of nature* is often thought of as essentially Romantic, though its cultural importance can be traced back to the period of the Enlightenment, to Rousseau in the eighteenth century and to Hobbes in the seventeenth century. Rousseau and Hobbes think of the state of nature as something that human beings have always already left behind – either as a state that long predates the advent of civilization and technology, or indeed as a hypothetical condition useful for gauging the overall direction of cultural development. If the civilization-wide catastrophes and environmental disasters of the twentieth century have taught us anything however, then it is that the state of nature is *not* something we can distance ourselves from historically or theoretically. We have learnt that we too remain natural beings and that we are dependent on a living natural environment. The learning process has indeed been largely negative. We have been alerted to our dependence on the natural world precisely as it has become clear that the natural ground that we have taken as an unquestionable given may not support us for much longer. The tremendous significance of artificial nature, its multifaceted role in contemporary life, says a great deal about the predicament that we find ourselves in – one in which we have come to accept that we live in nature and with nature, and yet one in which we can no longer count on natural life running its independent, self-sustaining course. And so we mourn what we have lost of the world of nature in the world of art. So we conjure a lost world of nature via simulacra, now seeking to repair and substitute for the nature that once was, now dreaming of new worlds designed to afford us a new sense of security.

Whatever else it suggests, the term 'artificial nature' at any rate signals a hesitation. It is something like a question mark hanging over a period of human history – our current era of technification – in which terminating the state of nature is almost a matter of explicit debate. Two paths look to be open to us, at least theoretically.[27] On the first path, human beings come to accept that they are natural beings and attempt to bring technological development into approximate harmony with their natural state. (It should be noted that going down this path does not mean abandoning technological development – that it would leave broad scope, for instance, for what Ernst Bloch calls *technical symbiosis*.) On the second path, humanity wagers on creating some sort of super-technology, i.e. on a large-scale technological breakthrough that allows us to definitively bid goodbye to the state of nature and take our existence wholly into our own hands. Both future pathways are already coming into view. However, which of the paths humanity will end up travelling is probably less a question of how technology itself develops than a matter of how we confront technology and nature in the domain of political debate and political action.

Nature in the age of mechanical reproduction

Introduction

In 1936, Walter Benjamin published an essay that was to fundamentally shape subsequent thinking about the difference between classical and modern art, particularly since its appearance in German in 1955. That essay is 'Das Kunstwerk im Zeitalter seiner technischen Reproduzierbarkeit', known in English under the slightly truncated title 'The Work of Art in the Age of Mechanical Reproduction'.[28]

In the essay, Benjamin notes that the world of modern art has been shaken to the core by the fact that art has become technically reproducible (he is thinking primarily of visual art). Nor is it just that the details – the materials, the subject matter or the techniques – of artistic production have changed, not just, for instance, that the whole development of modern painting is incomprehensible without reference to the competition that arrives with the invention of photography. Rather, *what an artwork is* has changed, and not just the new art of new generations of artists, but the entire corpus of existing artworks that come down to us from history, the very greatest included.

Benjamin doesn't specify all of the techniques of reproduction that he thinks his argument applies to. No doubt, he has in mind the technical forms that were increasingly available in the period when he was writing – photography, film and gramophone recordings. One thing, however, is clear – that, in talking about *technical* reproducibility, he is drawing a line between a new species of art and the sort of art made possible by manual reproducibility within the crafts. The focus of his attention is *technically perfected* reproduction – processes that allow not just for the fabrication of further examples of a general type of artwork, but for the *repetition of the individual artwork itself*. What the technologies of exact reproduction strike a blow at, Benjamin says, is artistic *authenticity*, the uniqueness and unrepeatability of the original work and the sense of aesthetic dignity that flows from that uniqueness and unrepeatability. Writing in 1936, Benjamin may have had an inkling of the destruction that was soon to be wreaked on a vast number of the originals of

European art and architecture. In the period after 1945, we have got used to seeing reproductions in the place of originals. Yet before that situation took hold, the mere possibility of reproduction had already led, in Benjamin's view, to a loss of the aura[29] surrounding works of art, to a dwindling, in other words, of that nebulous quasi-religious halo enveloping premodern art, which guaranteed the established artistic classics the status of historical individuals and, in turn, commanded a certain sort of respect and established a certain sort of distance between viewer and artwork.

The transformation that has taken place has made no immediate difference to *individual artworks*, for an aesthetic aura, as Benjamin defines it, is not something material that attaches itself to the individual material work. Rather, the loss of aura is one aspect of the *transformation of art as a social institution*, a change in the very social function of art which accompanies the rise of industrialized artistic production. Our very way of seeing art has been fundamentally altered by its mass reproduction, perhaps by the very possibility of reproduction. Benjamin speaks here of a certain process of displacement: in the later stages of artistic modernity, the cult value of works of art, he thinks, is converted into exhibition-value; increasingly, art's key function is its quite tangible role in the creation of a certain sort of social spectacle.

It is important to note here that the change to art as an institution, which Benjamin links to technological reproducibility, was something hastened by the artists of the time, above all by the *avant garde* of artistic modernism. Modernism pressed quite deliberately for the dissolution of art; one of its leading ideas was that art could be made to pass over into the broader stream of life. Many of the methods with which it pursued this goal are already identified by Benjamin; many that Benjamin doesn't mention, such as the deliberate laying bare of the fabricated nature of the artwork, would fit easily into the framework of his analysis; all aim at dispelling the aura of the artwork. What was often overlooked by modernism's *avant garde* – what Benjamin, however, is quite explicit about – is that modernism's demystifying strategies are in no way likely to lead to the abolition of art *per se*, but simply to the overhauling of more traditional concepts and the rise of a new concept of art. Benjamin's essay sets out to show that the techniques of reproduction that bring about the death of the unique artistic original and hence the end of the artwork in the classic sense also come to function as techniques for *producing* art of a new and different kind. Benjamin is clearly thinking of artistic products that exclude the very notion of an incomparable original because what is produced cannot but be presented mechanically by means of technical devices: art whose very essence as art depends on the existence of technologies of reproduction (the most obvious example would be film). However, confronted with the way such new aesthetic forms are supplanting the art of old, Benjamin refuses to strike up any sort of note of cultural pessimism. The 1936 essay connects technological reproducibility with the hope that art will escape the narrow confines of professional criticism and educated middle-class life. For Benjamin, if art is to have a future meaning, it is as part of a genuine *mass* culture. And, as he also clearly realized, what that means is that art will inevitably enter the arena of politics, either, he thinks, through the aestheticization of politics in fascism or through the politicization of art in communism.

Today, it is surely impossible to read Benjamin's 'Work of Art in the Age of Mechanical Reproduction' without agreeing *and* disagreeing with its core argument, and if our reactions are inevitably divided, this is partly because the social and political alternatives that we face are different from those facing Benjamin. Can we read the original essay without being affected by something of that mood of horror that Heidegger suggested[30] was the basic mood of twentieth-century philosophy? Any perceptive reader will sense that 'The Work of Art in the Age of Mechanical Reproduction' has something profoundly true to say, in its own way, about our entire historical situation, something that goes far beyond any sort of debate about the history of art. In the present essay, we want to bring out the broad – the very broad – truth of Benjamin's position by pursuing the same train of thought along a very different pathway. For experiment's sake, let us substitute the phrase 'work of art' in the title of Benjamin's essay with the word '*nature*'. Or rather, let us simply remind ourselves that art and nature have formed a conceptual pair since Western culture began with the ancient Greeks, and that a fundamental shift in the meaning of one of the two terms implies an inevitable, and probably a major, shift in the meaning of the other.

The loss of aura: for Benjamin that meant the elimination of that pathos of distance between work and viewer that was so conducive in the past to a sense of aesthetic awe; it meant the gradual destruction of all sense of aesthetic uniqueness, the loosening of traditional bonds and conventional associations, and the stringent reappraisal of art in terms of use- or exchange-value. In the domain of nature, loss of aura would appear to mean something quite analogous – a loss of all sense of the *naturally given*, a loss of respect for natural life, the destruction of all sense of the uniqueness of natural environments and species (of all sense of their existence as once-off phenomena that can be irretrievably lost), above all the systematic utilization of nature as a product. To one side of all romanticism and all easy sense of human affinity with nature, what ought to unsettle us about the loss of aura that so marks our historical predicament is the recognition that, in talking about nature, we are talking about ourselves as well: *the nature that we ourselves are* enters just as surely into the radical devaluation of nature as does external nature or the planetary environment at large. In short, the question posed by the technological reproducibility of nature places a question mark over the nature of humanity, and human beings' very conception of themselves.

Hubris?

Yet is the technological reproduction of nature not something we rightly shy from speaking of? Is the image of humankind as a second creator of the world of nature not a highly presumptuous one? Is the thought that nature is something that can be *mastered* – to the point where its continued existence is something that human beings are in a position to technically contrive – not in itself an expression of hubris? Is nature not that all-encompassing power whose operation we have to presuppose, and in some sense *should* presuppose, in everything we do? Nature (reputedly) always operates on a scale that is beyond human beings. Every improvement that we make to natural yields (e.g. crop yields), every act of destruction that we unleash on ecosystems and habitats, is something that can always be placed within a wider natural frame. But does that mean that what human beings do to the natural world – what they *make of* the natural world – is irrelevant? Evidently not. It seems that we are going to have to find the right scale to gauge human action vis-à-vis nature in order to begin talking about the predicament that we find ourselves in.

If we think of nature today unavoidably as a social or historical product of human activity, the nature we have in mind is, of course, the natural world here on earth, what used to be called the *sublunar* sphere; we mean nature as it exists on an order of magnitude that makes sense in terms of the scope and duration of human life. Nature here means nature in its relevance *for us*, however inadequate this notion of a nature that exists 'for us' might be.[31] Clearly, the human manipulation of nature is never without limits, never, we should say, unconditional; it always depends in one way or another on natural givens (a supply of basic natural materials for example) and it always has to make use of already existing laws of nature. Yet as the human interventions made possible by modern technology have continued to escalate, what has to be taken as a natural given has become increasingly vague. Once it was natural materials like wood and metal that were shaped by the manual labour of craftsmen. Then it was more complex elements that could be arranged and rearranged to form synthetic substances, such as ceramics, amalgams, crystals. In very recent history, we have moved on to elementary particles and atomic fragments, which can be split apart and fused together to form new elements on demand or hybridized to create isotopes of various kinds. Something similar applies to the *laws* of nature. While human beings remain subject to some very basic physical conditions – the law of gravity, the principle of the conservation of energy and various causal interactions – beyond that we can arrange things at will. In the highly synthetic environment of a nature shaped by human purposes, the greater part of regular causally determined action is for human beings to direct as they see fit.

How basic in practice is human action to the reproduction of the natural world and how far might it extend in principle? Let us start by surveying some of the major dimensions of human activity with this in mind, giving ourselves a little science-fictional licence in the hope that it will help us to see where our current rough course might be taking us. Two of the most ancient forms of human activity are worth considering first – agriculture and forestry. Here we can point to a fundamental historical break that takes place in the nineteenth century: Liebig's invention of artificial fertilizers and the advent of industrialized farming. Since Liebig, we can say, agriculture is no longer primarily a natural cycle of reproduction steered to their own advantage by human beings, but the formal reproduction of one particular natural function (a more or less specific type of fertility) with the help of additives obtained from far beyond regional natural cycles. In broad historical terms, the trend seems to be towards environmentally non-specific farming practices, in effect the development of large-scale agricultural factories that transform precise quantities of nutrients into biomass of a desired kind, located on land devoid of all naturally occurring vegetation.

Second, let us consider living creatures, or, more specifically, living species. The briefest of historical surveys ought to make clear what sort of shift is underway in today's world. The Greeks, after all, regarded species as eternal and unchanging, while in Christianity the natural divisions between species are creations of God: the ancient conceptual universe is organized in one way or another around a vision of nature as *the given*. As long as human beings have been in existence, they have bred animals and cultivated plant and crop strains to suit themselves, including many that would be incapable of reproducing themselves without human assistance. And while the theory of evolution tells us that biological species are subject to change, it also makes clear that the earth's given complement of species is just that – *a natural given*. Yet the ground marked out by Darwinian natural selection would appear to be shifting. Today human beings are setting about *productively* intervening in natural biological evolution, not just guiding natural reproduction by selecting the most useful variations from among different strains of biological life, but by deliberately producing and reproducing variations according to plan. The results are relatively well publicized: living beings designed according to functional specification and patented for use by their designers.

Third, there are what we conventionally call ecosystems: localized networks of organic and inorganic processes that are reproduced cyclically and modified over long periods in the course of evolutionary change. On this score, it has been clear for many decades that ecosystems are something like ideal types. Ecologists know that natural conditions no longer reproduce themselves spontaneously without the intentional or inadvertent help of human beings, either at the regional or global level; talk of ecosystems has been replaced by more realistic notions of 'human-organized ecosystems'[32] or, still more modestly, human-influenced eco-*structures*.[33] The preservation of highly valued natural habitats or regional natural landscapes requires ever greater labour on the part of human beings, ever greater supplies of external energy, the ongoing substitution of naturally occurring materials and an increasing degree of system-wide regulation. Nor is it difficult to imagine a utopian outcome of such a line of development, a system of perfect computer-guided eco-management that is capable of stabilizing natural environments rather than leaving them at the mercy of nature's faltering powers of self-regulation. The excessive demands that human beings are placing on the natural world have already resulted in attempts to reproduce nature in a quite different sense, by restoring damaged mechanisms of natural self-regulation. Environmental conservation, in this sense, passes over into environmental restitution. Measures undertaken, particularly in the developed Western world, range from the restoration of waterways to the cultivation of marshes and swamps – with their entire complements of microfauna and macrofauna – all the way up to the reconstruction of entire landscapes devastated by industrial overuse.

Fourthly, we have the purely aesthetic reproduction of nature, the reproduction of nature as it were *without nature*, in appearance only. A variety of cultural forms and (sometimes frankly bizarre) subcultural products come under this heading too, everything from drip-fed trees in concrete shopping malls to mass-produced plastic Christmas trees. The sort of aesthetic paraphernalia that we have in mind here aims at satisfying a quite basic human need for nature, even if in an obviously residual or mutilated form, by what we could call a *gesture of naturalness*.[34]

Next, we ought to mention human beings themselves. Human bodily nature, as we have suggested, is caught up in the problematics of the reproduction of nature as much as any other aspect of nature. While it was once one of the by-products of love and communally organized meals, human self-reproduction tends increasingly to cast off its previous cultural forms; it has now been taken in hand technologically. Meals could be on the way to becoming the ritual consumption of aesthetically appealing 'food-related products',[35] with the basic nutritional content supplied before or after the fact by nutrient supplements or vitamin concentrates in tablet form. Producing offspring could likewise come to be largely separate from physical love. Sex itself could take place in sterile bodies and unproductive moments, with the physical fertilization of eggs occurring outside the body according to plan and *in vitro* testing and selection, together with a range of prenatal therapies, guaranteeing that children are everything that parents could possibly want them to be.[36]

This brings us lastly to medicine. While nature is proverbially the best doctor, and while common wisdom tells us that the physician's role is to help nature help itself, with the advent of modern scientific medicine, reproduction in the general sense of returning the body to health is again something that has been comprehensively taken in hand. The spectrum of techniques at our disposal for reproducing human bodily health takes in everything from organ transplants and prosthetic surgery through to technically guided chain reactions designed to replace the physical mechanisms of bodily regeneration. At the risk of turning themselves into a kind of total computer-guided prosthesis, human beings are increasingly setting out to reproduce their own nature by technological means.

Nature and technology

My aim, however, is not to exhaustively survey the consequences of the technological reproducibility of nature. In what follows, I want to avoid asking whether what is technically possible in principle is likely to be realized in practice, or indeed whether it should be realized. A wealth of contemporary discussions of these topics already exists; the dangers that lie in the technological reproduction of nature – the raft of ethical questions, the associated question of alternatives – have been canvassed at length, and will continue to be. Rather, what I want to concentrate on is the fact that something fundamental has *already* occurred, *irrespective* of how far humanity ends up exploiting the technological possibilities of reproducing nature; I want to try to show that what nature is for us and hence the entire existential predicament of human beings, who themselves are part of nature and of whom nature is an integral part, have both already been radically reinvented.

We could also put the point a different way. Long before the possibilities of technological reproduction had completely altered what nature was for us, as an essential part of a broad European cultural complex nature had already been subject to a devaluation of values, if not a wholesale destruction of values. The fact that nature has become something of a cultural cynosure precisely in the contemporary world is, in this sense, quite misleading. Today unsullied nature is the ideal of human leisure (i.e. of the tourist industry), an important part of the aesthetics of what we produce and consume and a vital element of political programmes. Nature or 'the natural' is indeed a predicate that can be attached to anything and everything. As an epithet for consumer goods, nature is meant to signal a kind of raw positivity (natural fibres, natural cosmetics, natural colourings), with the prefixes 'bio-' and 'organic' functioning as intensifiers (organic wine, organic produce, bio-energy, etc.).

Considered as a positive ethico-aesthetic value, nature, we might say, is a latter-day product of a long European tradition. As an ideal, the notion of nature was underwritten throughout European history in an ontological, cosmological or theological sense and stood in a relation of tension to the ideals of culture and civilization. Expressions like natural law, natural rights, natural (i.e. pristine) conditions and 'the nature of the case' testify to nature's normative role. Implicit in all such idioms is a conception of a *given* condition of all individual things and all human relations, an *original* order of the world, which by comparison with the world that we fabricate for ourselves is a world we can easily lose sight of.

Since the Greek Enlightenment of the fourth century BC, the conceptual contours of the term 'nature' have been shaped by a series of oppositions: nature has been understood in *contradistinction* to technology or culture or civilization or human rules. In each case, nature is thought of as *standing opposite* the domain of the human; in general, we might say that since the Greeks 'nature' denotes what exists of its own accord, what is self-originating, i.e. what spontaneously produces and reproduces itself, while the domain of the human is characterized by arbitrary stipulation, by law, by production of quite a different kind, and by the fact that what is, and what ought to be, has to be affirmed by human beings, established through human effort and reproduced through human agency. One of the most prominent members of the series of oppositions here is *physis/techne*. *Physis*, which gives us our word for physics, is in fact the Greek word for nature; etymologically, it is simply the term for anything that arises or appears of its own accord. *Techne*, on the other hand, is the collective term for a range of human capacities and for specifically *productive* knowledge; notably it refers to what we would today call technology, including craft tools, craft knowledge and agriculture, and to the realm of art. On the Aristotelian picture, *physis* is the realm of what exists of its own accord. A naturally existing being is something that has its principle of movement in itself.[37] Compare this with a technological entity, which for Aristotle is something whose principle of movement is imparted to it by human beings for human purposes – something that depends on human beings in order to come into existence and continue to exist. Aristotle's example of the distinction between the two is justifiably famous: if you bury a willow bedframe, what will grow out of the buried frame will be a willow tree, not a bedframe. The self-reproducing willow belongs to the realm of *physis*, the bedframe to *techne*.

In this quintessentially European conception of nature, nature is what exists of its own accord and is self-reproducing. As we have seen, for the Greeks nature, so conceived, has existed from eternity, while on the Christian interpretation it has existed since the moment of divine creation.

Now we can certainly say that this conception of nature has been in steady retreat since early modern times. Nature in its entirety was already pictured as a vast clockwork mechanism in the Renaissance. In the work of Descartes, we can even read the original programme for understanding nature in the light of that metaphor; on the Cartesian picture, nature is something that human beings seek to understand as a craftsman understands his craftwork, viz. as though it were something made or constructed. (The interpretative analogy actually runs deeper. In Descartes, all animal life, and the human body too, is to be conceived on the model of the machine.) Moreover, whether or not one agrees philosophically with Descartes, since his time the cosmic/ontological gulf opened out by the possibility of technically reproducing nature is something one must face up to: if we think of nature as a piece of clockwork or a kind of work of craft, then we are thinking of it as the work of a craftsman, or of some sort of divine eminence whose creative abilities *definitively* exceed the productive capacities of human beings.

Newton is known to have referred both the question of whether the solar system evinced systematic order, and, even more fundamentally, the question of the basic properties of matter, to theology: he considered both to be non-scientific. Kant, for whom Newton's work provided the model of true scientific explanation, followed him; for Kant, Newton's high ideal of a comprehensive explanation could never be applied at the level of nature's individual works (there would never be a 'Newton for a blade of grass' as he put it). Justus von Liebig, that impassioned nineteenth-century opponent of the romantic philosophy of nature and an outspoken advocate of experimental science, still thought it necessary to posit the existence of a *vital force* to explain the organic order of the natural world. The list of the great scientific minds who swore by some version of the classical conception of nature could be extended much further. Our contemporary scientific picture could hardly stand in more marked contrast to that of Newton, Kant, Liebig et al.; the splicing of DNA sequences, our quintessentially technicist understanding of elementary particles, our penchant for constructing new elements, and so much else, places us at a distance, not just from the ancients, but also from the greats of *modern* science. The quandary is not just that we no longer acknowledge any in-principle limits to what we can fabricate from basic natural materials. Nor, indeed, is it the fact that we are in the process of *literally* realizing the Cartesian programme of knowing nature from the standpoint of its constructability. Nature is simply no longer the given in any meaningful sense at all. In the course of the past century, it has become what it is in principle possible to make or fabricate.

It is relatively easy to see that this notion of nature as what can be fabricated is deeply paradoxical from the standpoint of the classical conception of nature. One might say such a notion reduces the classical conception to absurdity. Yet on the other hand, it would also seem absurd not to count elements with atomic numbers over 100 among the *natural* elements, or to refuse the predicate of the natural to human insulin produced by genetically modified bacteria rather than human beings. Likewise, a clear distinction between polymers manufactured naturally by plants and synthetic materials generated by man-made processes of polymerization has simply become impossible to draw.

It is difficult to overstate just how revolutionary a change nature has undergone in our time. The possibility of technologically reproducing nature signals the end of our past conception of nature, whose pathos lay in the contrast it enabled us to draw between nature and things of our own making. Nor is that rather broad shift without its relevance for particular debates about nature that bulk large in the public domain. The contemporary invocation of nature as an ideal value reveals itself as partly ideological, precisely in appealing to a conception of nature as something fixed and definite at the very moment at which that conception is breaking down in a dramatic and probably irreversible way.

Nature and art

The traditional opposition between *physis* and *techne* – nature and technology/art – does not just imply that the natural and the artificial are set off against each other, it also suggests an inner connection or mutual affinity of sorts. That affinity becomes especially clear in classical aesthetics. Starting with the art of the Greeks, nature and art cast a special sort of light on each other. As Kant puts it in a passage of his *Critique of Judgment*: 'Nature proved beautiful when it wore the appearance of art; and art can only be termed beautiful, where we are conscious of its being art, while yet it has the appearance of nature'.[38]

Kant's point here needs some explanation. To start with, we shouldn't be misled by his use of the past tense (nature *proved* beautiful, etc.). Kant speaks in this way purely in order to connect what he is saying to the discussions of previous commentators. His proposal is thus that we call nature beautiful when it has the appearance of art. Yet what does 'wearing the appearance of art' mean? Kant, in short, thinks that nature exhibits an order and a law-like regularity that we can only grasp conceptually by thinking of it *as if it had been produced according to a plan*, i.e. as if the world of nature were a product of conscious technological invention. The *Critique of Judgment* thus speaks of a *technic of nature*: an earlier section tells us, for instance, that 'self-subsisting natural beauty reveals to us a technic of nature'.[39] Again, what that means is that we can only grasp the notion of nature's essential character opened up to us in aesthetic experience if we conceive of nature on the model of technology, as something quasi-technological.

In the second part of our extract from §45 of the *Critique*, Kant comes at the point from the opposite direction as well; he tells us that a proper concept of art requires a concept of nature: '. . . art can only be termed beautiful, where we are conscious of its being art, while yet it has the appearance of nature'. (We note here that when Kant talks about *schöne Kunst* – literally *beautiful* art – he is talking about what in English is called *fine* art. He needs to do so to make it clear that he means something more specific than art as his contemporaries understood it, for until the formation of aesthetics as a discourse around Kant's time, the term 'art' would have been associated with the purpose-built products of craft and technology as much as with painting, poetry, music and the like.)

Now Kant's argument is that works of fine art are defined by the fact that the artifice involved in producing them is inconspicuous; nor is it just that we *don't* notice the artifice involved, we *shouldn't* notice it. Works of fine art give the appearance of existing naturally of their own accord, and so they should. For Kant this is the very essence of art, as it is for all those who share in the classical conception of art: to create a work of art is to deliberately give something that one produces the appearance of natural self-dependence, to endow something whose production conforms to rules with the appearance of purposelessness, spontaneity, effortlessness, etc. It is easy to see how the idea ties in with the concept of artistic genius that was soon to flourish in the European nineteenth century: the artist as genius becomes the human being in whom and through whom nature effortlessly and spontaneously produces itself.

Classical aesthetics thus suggests that, although nature and technology are conceptual opposites, and although their identifying marks are mutually exclusive, they are also concepts that implicitly refer to each other; in important cases, the concept of nature is something that we have to think through the concept of technology and vice versa. Nature becomes the object of our admiration precisely when it has a semblance of artistry. And art attains its proper condition as art when it leaves all mere technique behind and takes on a natural aspect. (To get the classical picture in full, we have to add a second moment to the thought: art is seen classically in the eighteenth century and for some of the nineteenth century as the *imitation* of nature – as, indeed, is technology until well into the twentieth century, as we have observed in the case of Kapp and Gehlen's respective theories of technology.)

The way that the notions of art and nature are interwoven in the European conceptual imagination ought to leave us in no doubt that the dissolution of the classical concept of art has left its mark on our concept of nature too. The long-standing interconnection between art and nature might suggest to some that the technological reproduction of nature need not represent such a troubling moment of rupture; after all, if nature has always been implicitly regarded as deeply admirable when it had the appearance of a conscious technical creation, should we not welcome contemporary efforts to create nature anew on a technological basis? That line of reasoning overlooks one important thing, viz. that in thinking through these issues, we find ourselves in the domain of art, which is to say in the realm of appearances. And in the realm of art, appearances are *of the essence*.

To put the point less paradoxically, in assessing the impact of the dissolution of classical aesthetic values, we need to take into account that the obliteration of appearances means the obliteration of the thing itself. Classically, it was all important that, despite appearances, art was *not*, in fact, something natural, and that nature was not, *in reality*, a technological mechanism. Modern artists' attempts to produce automatic art – either by letting the Unconscious speak directly or by creating various forms of art with the help of stochiastic mechanical processes – were among the most effective solvents of the classical aesthetic order. The way that modernist artists left behind visible traces of art's nature as artifice should also be mentioned here. All such tactics were part of a conscious effort to destroy art's appearance of quasi-natural self-dependence, indeed to destroy any impression that works of art exist in and for themselves. In Benjamin's language, we might say that the *aura* of art has been destroyed through a series of deliberate strategies of demystification, by consciously drawing attention to the technical basis of art, and by the deliberate creation of random aesthetic effects. In the process, the artist has lost his halo too, as Baudelaire's pretty parable intimates: in the whirl of street traffic, the aureole belonging to a poet landed one day in the gutter – and there it has stayed, despite the best efforts of subsequent generations of twentieth-century aesthetes. Or, as Musil was to put it a little later, in the course of the nineteenth and twentieth centuries, the poet became a mere wordsmith, the painter became a 'paintsmith' – and on it went.

Yet within the same period, nature too was to lose its subtle air of other-worldliness – that nimbus of sublimity that constituted its value as one of the guiding threads of European culture. What did the nimbus consist in? In retrospect, we can say that, as long as it remained in place, nature appeared to European man as the product of masterful technological artifice (although in reality it was no such thing) and that it did so because it exhibited order and regularity. Kant argues that human beings have a *moral* interest in the appearance that nature is something sublime and purpose-built, and that that appearance is simultaneously constitutive of our sense of the beauty of nature – it has been particularly manifest in the equation of the natural with the *good*, the *original* and the *innocent*, that has played an influential role in Western intellectual and cultural life since the eighteenth century. On this picture, the order of the natural world is praised as beautiful because it is an order that is quintessentially effortless; while civilized human beings have to exert palpable self-control in order to remain within the bounds of order and beauty, nature displays beauty and order of its own accord. The aura surrounding nature and natural beings is thus a correlate of the wonder that we feel when faced with the harmonious totality, the 'order' that nature seems to organize itself into, its 'purposiveness without a purpose', as Kant calls it. In classicism, it is right to say, the synthesis of aesthetics and ethics would appear to be complete: human beings react to the sublimity of the given order of nature with a sense of *moral intent*.

The aura of nature has been lost, however, not just because the works of nature have come to be seen as comprehensible on the model of works of technological artifice, but because they have, in point of fact, *become* works of technological artifice. It is this loss of natural aura and the co-ordinate loss of nature as a cultural ideal that we are confronted with as the human capacity for technical reproduction advances. However, in identifying what is occurring as a rapid process of demystification, I by no means want to suggest that modern natural science diminishes our sense of natural mystery or awe. On the contrary, the 'technics of nature' that patient scientific investigation has uncovered in recent decades, the sophisticated intercellular protein factories turned up by recent microbiology for instance, seem far *more* amazing than anything that the eighteenth century thought nature was capable of. In short, it is not scientific knowledge of nature, but the enormous step that we have taken towards *fabricating* the natural world in the literal sense, that leads to the loss of aura. As awe-inspiring as the process of scientific discovery continues to be, the whole experimental emphasis of contemporary science – which is often a prelude to the manipulation of natural processes – already *presupposes* a widespread loss of respect for natural life.[40]

Our ability to technically reproduce natural processes in short leads to a collapsing of the respectful distance between ourselves and external nature, though the fact that a definite line is here being overstepped is something we might only detect in the enthusiastic rush of publicity with which scientific progress is announced to the wider world.

In approaching the problem of nature and technology from an aesthetic direction, it seems appropriate to choose an aesthetic example of the loss of natural aura that we are discussing. It comes from the young science of fractal geometry. Recently, I was treated to a curious display of mathematical mimicry, a demonstration that copying three rectangles over the top of one another with constant reductions to scale is enough to generate a strikingly life-like image of a fern leaf. Now, the fact that it is possible to take a simple geometrical form and produce a highly refined and indeed highly *unusual* natural form using a single recursive formula will probably bring a little joy to the heart of everyone schooled at some stage in natural science. Nonetheless, the total effect of the demonstration was unnerving. A cultural influence from a different quarter is clearly at the root of the shock that we feel at the ease with which a paradigmatically complex natural form can be reproduced once it has been reduced to schematic simplicity.

Kant supplies us with a second example of the loss of aura:

What do poets set more store on than the nightingale's bewitching and beautiful note, in a lonely thicket on a still summer evening by the soft light of the moon? And yet we have instances of how, where no such songster was to be found, a jovial host has played a trick on the guests with him on a visit to enjoy the country air, and has done so to their huge satisfaction, by hiding in a thicket a rogue of a youth who (with a reed or rush in his mouth) knew how to produce this note so as to hit off nature to perfection. But the instant one realises that it is all a fraud no one will long endure listening to this song that before was regarded as so attractive . . . It must be nature, or be mistaken by us for nature, to enable us to take an immediate *interest* in the beautiful as such.[41]

Kant's anecdote is – by Kant's standards – a funny one, so it fails to really bring out the serious side of the loss of aura, perhaps in part because the scenario that Kant sketches presupposes a clear and fixed distinction between the actual phenomena of nature and *what we take to be* the phenomena of nature. In fact, part of our contemporary sense of unease vis-à-vis nature stems from the fact that what science confronts us with is the increasing volatility of this very distinction.

The technological reproducibility of nature divests nature of its aura. Moreover, it makes nature redundant as a focal point of cultural experience. European culture has historically taken the concept of nature as a point of orientation, a keystone of ethical self-understanding, a stable pole of legal, social and political thought. It is beyond the scope of the present work to trace the specific repercussions of the dissolution of the classical conception within law, politics and the wider social domain. In what follows, I have chosen one area where these issues play themselves out to particularly serious effect – the domain of human self-understanding. Nature, there can be no doubt, is at the very heart of the issue of the self, for *the nature we ourselves are*, the instinctive bodily existence that is the most intimate possession of every human being, was historically one of the fixed points of departure for self-understanding, self-experience and self-definition.

The nature we ourselves are

Kant separates philosophical anthropology into two distinct parts: physical anthropology and what he calls anthropology from a pragmatic point of view. Pragmatic anthropology for Kant addresses the question of what human beings can make of themselves and what they *ought to* make of themselves. This he contrasts with physical anthropology, which addresses what makes up the human *physis* – the naturally given aspect of human life. Human nature in the sense of *physis* is the nature we ourselves are: the body.

Until relatively recently in history, everything connected with this bodily nature that we ourselves are, including everything that grew out of it, was accepted almost as a brute fact; integrating the givens of bodily existence into the individual's life project meant taking a particular attitude towards them, at most instrumentalizing them to one's advantage. One was born as a man or a woman, with a particular constitution and particular dispositions, perhaps with certain illnesses as well. The body could, as it were, strike one down with disease; instincts and moods that one struggled and engaged with welled up from within the body, and something similar applied to what literally grew out of the body: one was either blessed or burdened with children, in either case they were something that *appeared* in one's life and that one impassively accepted. Body, bodily constitution, illness, children – these were what primarily if not exclusively constituted one's fate, which was itself a kind of superordinating given condition that the individual had to struggle and engage with. Yet the confrontation with the givens of bodily experience also represented the wellspring and strength of the self. Moods and drives were a spur to self-assertion, occasions for self-mastery and the formation of character.

What will become of human beings when what once constituted their basic bodily nature finds its way into the list of technologically reproducible contingencies? What will constitute the future basis of human dignity and self-respect if the boundaries between the domains of subjective self-determination and the naturally given continue to fade? When the difference between what Heidegger calls *faciticity* and *project*[42] becomes even less stable than it now is?

Human beings up until our own time were, in part, defined by the possibility of *being affected* by the vicissitudes of bodily existence, or, as one philosopher has put it, by the fact that, in addition to having a body, I also *am* my body.[43] Yet how is the positive capacity for being affected, this *sensitivity* that is so essential to the development of human *sensibility*, to be developed and how is the acknowledgement that I *am* my body to be maintained if all bodily organs, including the heart, can in theory be replaced? How is unambiguous self-identification with a particular gender to remain a touchstone of self-understanding if my gender is something I can change at will by means of hormone therapy or surgery? Can we speak as we once did of the formation of the self or of individual self-determination when we are no longer obliged to struggle or engage with our individual physical constitution, drives or moods? When the possibilities of struggle or engagement are no longer a question of will, but a matter of adroit rational control or the effective use of medication? Is the burden that children represent going to be something that parents are existentially equal to when children are no longer something one passively accepts into one's life, i.e. when responsibility for their very existence falls wholly and solely to parents (e.g. when it depends specifically on their use or non-use of contraception)? How are parents to deal with the unavoidable problems posed by children's gender, bodily constitution and bodily infirmities if prenatal testing, selection and therapy also make the basic facts about a child's bodily being squarely into matters of parental decision – and parental responsibility?

The possibilities of technologically manipulating physical reproduction have led some theorists to observe that human nature itself is in the process of becoming contingent.[44] We can put the point in even sharper terms: insofar as nature meant the factually given, the *données* of fate, etc., human nature is in the process of totally dissolving. The question is very much what that means specifically. One thing it doesn't mean is that there is no longer anything fate-like or subject to chance, no longer any *givens* of human bodily existence. The tension between what Heidegger thinks of as *faciticity* and *project* is something that all human beings will feel the pull and counterpull of for all future time. The key point is that the position of the boundary between facticity and project is now something for human beings to decide: from our moment of human history onwards, what we can take for granted, or must take for granted, as brute fact will *itself be a human project*. Here we perhaps glimpse the depths of the changes that the technological reproducibility of nature has generated, as well as the irreversibility of those changes. The dissolution of nature (nature in the sense of the givens of human *physis*) has nothing at all to do with whether human beings *individually* choose to subject their bodies or their offspring to technological manipulation or not. We can put it this way: someone who chooses *against* regulating his body by using whatever medical paraphernalia may be necessary to keep him fit and active will still be expected by society to perform at a level comparable to that of others who make the opposite choice. Or, to give an example where the problem becomes even more acute: parents who decide against prenatal testing can no longer accept a child born with a severe disability as a simple given of fate. In short, now that the technological means exist, we each have to decide where we will draw the boundary of the natural.[45] The disturbing point is this: even what we leave in nature's hands, i.e. what we choose not to intervene to change, is no longer the given in any simple sense. It is certainly not the given of human *physis* as that has been understood in the past, viz. as something that, in being given to us, was also presented to us as a task and a responsibility.

Conclusion

When we reflect on the broad consequences of technological reproduction on the analogy of Benjamin's 1936 piece, the conclusion seems unavoidable that we are destroying nature as it once was, in the same way that we have destroyed art. Yet that verdict requires substantial qualification: when we speak of destruction here, we are not talking primarily about the devastation of nature in the concrete sense, but about the devastation of nature as a cultural idea, the devaluation of nature as a point of orientation within Western culture. In short, we are in a situation that is directly analogous to the situation brought to light in Benjamin's essay: just as the death of art pronounced by the Dadaists and surrealists has not meant the end of art per se, so the physical devastation of the natural world does not imply the destruction of nature per se, nor does the dissolution of the classical concept of nature imply the demise of our entire conception of nature. New technical processes of reproduction have led to the development of new forms of art, different forms of art. The question of what art will be in the age of mechanical reproducibility is still being answered. The question of what nature will be in the age of mechanical reproducibility can be placed directly alongside it: it too is an open question.

However, at this point, interpretative paths diverge. One can only extend Benjamin's reading of aesthetic developments so far in attempting to answer the question what nature is in the process of becoming. The hopes that Benjamin placed in the new art of the technological age have proved to be illusory in almost every respect. Mechanically reproducible art has failed to bring a proletarian mass culture into being,[46] just as the politicization of art failed to produce any significant results that one could genuinely welcome.[47]

With this in mind, we can draw the following conclusions. The technological reproducibility of nature has deprived our conception of nature of the definite outlines it once had because of the way that nature was defined in opposition to culture, civilization, humanity, etc. That means that, insofar as nature is a matter of practical relevance for us, it will, in future, have to be understood as *itself a product of culture*.[48] Yet it also means that humanity will come to see itself increasingly as belonging to nature; our cultural and technological capacities look set to be included within the scope of a new and much wider definition of the natural. In the process, the concept of nature will lose its definite contours. In a certain sense, it will return to being a concept of totality, as it once was for the pre-Socratics. That will open out the possibility of a new philosophy of nature – the philosophy of nature understood as *first philosophy*. Yet it will at the same time close off the possibility of deriving moral norms from the concept of nature. Normative distinctions will have to be drawn *within* the concept of nature as the possibilities of distinguishing things *from nature* fade. That too will have dramatic intellectual and existential consequences. In cultural terms, it will doubtless mean taking an *even more radically sober view of ourselves* than the one forced on us by the fading of so many of our other ideal values. Whether the new more comprehensive concept of nature that we see coming together around us today is also capable of disclosing more hopeful perspectives is difficult to say.

If the hopes that Benjamin placed in the technological age's new concept of art proved to be illusory, this was because the political alternative that he took his bearings from, the alternative between fascism and communism, actually obscured the fundamental historical developments within modernity that were taking place at the time, and continue down to today. Until recently, our gaze, too, has been diverted from the main course of development by a related dichotomy: the conflict between the Eastern and Western blocs representing the so-called communist and free worlds. It is only now that the Cold War is over that we can see clearly what has long been taking shape: a situation in which the anthropogenically created natural environment in which humanity lives, mediated by a conflict between the geographic North and South, will make a decisive difference to the course and indeed the outcome of all major social and political change. Changes to climate, as well as to the habitability and fertility of the earth, are on their way to becoming political issues of the first order for future society. The politicization of nature, there can be no doubt, is just beginning.

6

Critique of technology

Guided by an interest in rational conditions

Foundations of critical theory

In many ways, Max Horkheimer's well-known essay 'Traditional and Critical Theory'[1] lays the foundation stone of critical theory, and while Horkheimer's text and Herbert Marcuse's reply[2] leave one in no doubt that the original group of thinkers associated with the Frankfurt Institute for Social Research saw critical theory as something that had been in existence for many years – in practice since Marx's *Critique of Political Economy* – nonetheless, Horkheimer's essay is where the critical theoretic enterprise attains theoretical self-consciousness for the first time. In his essay, Horkheimer defines critical theory in contrast to a theoretical type he calls traditional theory, then outlines his own theoretical innovation in terms of a particular set of formal features and a particular methodology.

This attempt to characterize critical theory in scientific terms – we will sketch it in full in what follows – is of immense importance for one main reason, viz. because it enables us to detach critical theory as a scientific enterprise from the tradition that was established by its founders and first practitioners. From this point on, critical theory need no longer be identified with the philosophy of Horkheimer and Adorno or their various students, and that, in turn, should make it possible to extend its range, indeed to give it new life in the present – for instance, by applying its methods precisely to topics and fields like technology and nature that were far from the centrepoint of interest for the theory's founding fathers.[3]

Yet an obstacle stands in our way here, something alluded to in the phrase of Horkheimer's that provides the title of the present essay. Let me highlight that we are dealing with one of the defining moments of the critical theoretic project by quoting the full sentence from 'Traditional and Critical Theory' in which this phrase occurs: 'In fact, however, the self-knowledge of present-day man is not the mathematical knowledge of a natural science with the appearance of an eternal *logos*, but a critical theory of society as it is, a theory dominated at every turn by an interest in rational conditions of life'.[4] Critical theory, Horkheimer says crucially, is 'guided through and through by an interest in rational conditions'. Its capacity for criticism follows from having at its disposal a concept of what rational social conditions are, or at least a preconception that further reflection aims to refine. The question is whether a detailed picture of what rational conditions are, or, say, a concept of rationality itself, should be considered a necessary precondition of critical theory. Is such a concept of rationality to function as a sort of horizon within which – and only within which – the criticism of existing conditions unfolds? Should we expect the specifics of this type of theorization to be justified in terms of some such concept of rationality? Unless we take questions like these seriously, the possibility of extending and revitalizing critical theory might prove to be largely fanciful. And, indeed, the questions become increasingly urgent the more one notes in Horkheimer's text a striking absence of doubt about what rational conditions amount to, at least in the era that he is writing in; what Horkheimer has in mind are none other than the well-known lineaments of Marx's conception of society under socialism, above all free association among producers and appropriation of the forces of production in the interest of society as a whole. And while those two items may not take in the full range of rational conditions because *free* association clearly needs to allow for human beings to freely determine their own relations to each other, the limiting conditions imposed by such a rational framework are relatively obvious.

Today, the question we have to ask is whether the project of critical theory is viable without Horkheimer's specific picture of rational conditions. A series of open questions presents itself to us in the contemporary world that Horkheimer's presuppositions at least partly foreclose. What, we might ask, would rational technology look like today? What about a rational approach to the natural world? How is a notion of rational conditions to guide the development of new technologies, the use of existing technology and the interpretation of technology as an institution of human life? What would constitute rational human relations with nature, or, in other words, what kind of relationship with external nature and what kind of relationship with the nature we ourselves are would follow from an adequate contemporary notion of rationality? Before turning to this troublesome nest of problems though, let us take a closer look at the origins of critical theory, even if that means making the task of answering our guiding questions more difficult.

The Marxist paradigm: The critique of political economy

As we have said, the paradigm of the original critical theoretic project is Marx's *Critique of Political Economy*. Horkheimer's critical theory sees itself as an historically preconditioned extension of Marxist theory. What, then, defines critical theory as a theoretical move beyond Marx? On Horkheimer's account, what is essential here is an intellectual confrontation with *traditional theory* and, in particular, a sceptical stance vis-à-vis traditional theory's definitive instantiation, viz. natural science. The peculiarities of critical theory arise because it has a different object to natural science. Whereas natural science constitutes itself as an investigation of *nature*, critical theory constitutes itself as an investigation of *society*. Critical theory, in Horkheimer's lapidary formula, is the self-reflection of society. Yet society is an object of a peculiar kind: first, it is self-organizing, and second, it is capable, at least in principle, of organizing itself in accordance with an overarching idea of rational conditions. The theory sets up a conceptual opposition between nature and society, and indeed the fact that it does so was to have far-reaching consequences as late as the so-called positivist controversy within German sociology in the 1960s.[5] The battle here broke out because one party to the debate was looking to model itself on natural science and conduct research along positivist lines. Horkheimer and his followers were ranged against them. What followed, as it were by default, from the very structure of the dispute, was that Horkheimer simply assented to the positivist discourse of nature that found practical expression within natural science. And, indeed, the entire critical theoretic tradition followed suit. If the proper object of critical theoretic research was society, then the inner workings of a science whose object was nature could be left to one side. Critical theory continued to assume implicitly that natural science was a science of brute facts and was impracticable on any other basis. It overlooked, one might say wilfully, something that Marx himself had already begun to think through,[6] viz. that the development of the forces of production makes *nature itself* increasingly into a constitutive moment of social reproduction. Nature, in short, is increasingly becoming a social product and that, in turn, calls for a theory of *socialized* natural conditions. To put it in the language of what we have called *social natural science*,[7] what we need is a science of socially constituted nature, and that means, in effect, a critical theory of nature that is well beyond the scope of critical theory as Horkheimer conceived it.

Horkheimer's crucial oversight on the topic of socially constituted nature – though in a sense it is a simple reflection of faith in the intellectual foundations of natural science – has parallels too in critical theory's approach to technological development. The problem here has its roots in something other than an easy conflation of natural science and technology; it is rather that critical theorists have tended to view technological development, and hence also the development of the forces of production, as an independent variable within the wider framework of social development. Horkheimer thinks that the state of development of existing productive forces in some sense materially underwrites his regulative ideal of rational conditions, ensuring that the latter is more than a mere abstract utopia. As he puts it in the original essay: 'Such an association is not an abstract Utopia, for the possibility in question can be shown to be real even with productive forces at their present stage.[8] And Marcuse's view at the time was largely the same, though at one stage of his reply one begins to see why it would be him rather than Horkheimer who would later raise the possibility of a critical theory of technology and a critical theory of nature: 'There is no unbroken trajectory of progress from the old to the new forms of society, even as technology develops and the forces of production unfold', according to Marcuse. Technological development, in other words, cannot be relied on to smooth the path that leads from capitalism to socialism.

It was the opposite assumption, the opposite *hope* indeed – viz. that the development of the forces of production would provide the decisive means for creating Marx's realm of freedom, which pushed critical theory towards technological positivism, as it was also to do in the world of 'real socialism' in Eastern Europe. Critical theory tended to accept technology as it was, in large part because it was such a vital presupposition of its notion of rational social conditions. Had it followed one or two of Marx's own leads in different directions or run further with the work of Edgar Zilsel and Henryk Grossmann, it might indeed have thought its way beyond this relative technological naivety.

As things turned out, the School only shook free of its initial assumptions at the point when it was no longer possible to ignore the equivocal course of twentieth-century technological development. The negative effects (and side effects) of technological development were indeed to eventually lead critical theorists to ask how technology came to be constituted as a social structure. In other words, criticism of technology was *followed by* the question of a critical theory of technological development. This in itself was to create the theoretical difficulties that we have already referred to: how, for instance, is a critical theory of technological development to come into being in the absence of some sort of prior notion of *rational technology*? Although it is perfectly clear that a satisfactory theoretical answer to that question cannot follow any *aprioristic* logic implicit in technology itself, it is equally obvious that the theoretical challenge has not been met by more recent attempts to dialectically integrate the negative consequences of technological development on the analogy of the *Dialectic of Enlightenment*.

Another facet of critical theory that Horkheimer elaborates on in contradistinction to traditional theory is essential to understanding the current state of theoretical play. As the Greek provenance of the concept of *theory* suggests, traditional theory is essentially knowledge of *what is*. Critical theory, on the other hand, as Horkheimer explains, is oriented in a fundamental sense to *the future*. What it considers rational conditions are not just to function as a yardstick of social or cultural criticism, but as *goals* that are to be realized, in part through critical theorization itself. Horkheimer insists that any act of theory also has an inalienable social function; however, while the social function of natural science remains foreign to the latter's conception of itself, the critical theorist lays hold of his theory's social function quite explicitly – that function, indeed, becomes a constitutive moment of the theory itself. The 1937 essay thus speaks of a *goal* that critical theory exists to serve, in negative terms emancipation from inequitable forms of domination, in positive terms the creation of rational social conditions. In effect, critical theoretic practice takes precedence over theorization. In Horkheimer's own words, critical theory is 'the intellectual side of the historical process of emancipation',[9] where by emancipation we are to understand the emancipation of the working class.

The anticipations within critical theory of what rational conditions would look like are not a blank normative screen against which social conditions as they stand can be critically mirrored. Instead, they supply a practical purpose that critical theory itself exists to serve. Yet again however, we seem to come up against a weakpoint of the theory in its canonical form, for this suggests just how problematic it is to expect that the formal qualities of critical theory – i.e. the very thing that distinguishes it as a serious contribution to scientific knowledge – could ever gain broader legitimacy without being integrated into a political practice explicitly aimed at producing rational conditions. The formal quality I have in mind above all is the theory's characteristic use of paired concepts, which Horkheimer adverts to indirectly in discussing Marx's *Critique of Political Economy*: 'The Marxist categories of class, exploitation, surplus value, profit, pauperization and breakdown are elements in a conceptual whole, and the meaning of this whole is to be sought not in the preservation of contemporary society, but in its transformation into the right kind of society'.[10] Each of the individual concepts that Horkheimer mentions here – to make the thought entirely clear – unlocks its critical potential in dialectical *tension* with a related but radically different sister concept. Thus, the concept of the reproduction of labour-power stands in dialectical relationship to the concept of exploitation, value is the conceptual flipside of surplus value, a product's price is the flipside of its subjective value to the consumer, the surplus value created by the worker is the flipside of the profit extracted by the capitalist, and so on. However, in the context of the *Critique of Political Economy*, these deliberately equivocal conceptual pairings do not just enable Marx to pin down the operations of the capitalist economy; they also allow him to criticize existing conditions with a view to the creation of rational conditions *in such a way as to make the creation of such rational conditions appear politically achievable*. The analysis depends most famously on Marx's identification of a self-destructive tendency within the capitalist system that is also essential to its workings. Marx stakes his claim to be a truly revolutionary theorist on the transformation of the proletariat into a revolutionary force: the theory is predicated on capitalist 'pauperization' (impoverishment) leading to the development of class consciousness and thence to revolution.

The more one realizes how deeply embedded Horkheimer's interpretation of critical theory is in the theoretical context of Marx's *Critique of Political Economy*, the more difficult it becomes to reinterpret critical theory in light of contemporary social conditions. The main difficulty here arises from the historical experience of putting Marxist theory into political effect – a topic that has been canvassed by so many latter-day interpreters that we will leave it aside. That problem is compounded though by a second-order problem that we now need to address, the problem of specifying, even in the most basic of terms, what rational social conditions would actually amount to.

What is rational?

My argument is essentially this: what separates us in the contemporary world from the critical theory of the 1930s and the 1940s is the lack of a concrete picture of reason, or, in other words, a picture of what rational conditions look like. The question that follows from that predicament is whether reconstructing a picture of rational conditions is necessary if we are going to meaningfully practise critical theory at all in the present?

Habermas' *neue Unübersichtlichkeit* – his general term for the new and radical difficulty of gaining a perspective on society as a whole – has been put forward in the period between 1984 and the present day as a general characterization of post-modernity itself. Indeed, Wolfgang Welsch has gone one better in a lengthy volume, *Reason*,[11] and dubbed the same phenomenon *Unordentlichkeit*, a new *disorder-liness*. An important point seems to have gone astray, however, in discussions of these and related concepts, viz. that in Habermas' original Spanish lecture in 1984, *Unübersichtlichkeit* had nothing at all to do with any supposed postmodern tendency towards pluralism, but with the disintegration of a concrete notion of rational social conditions. In the original lecture, Habermas did indeed turn his attention in passing to critiques of science and technology (critiques that in subsequent decades have become a mainstay of critical theoretic thought); however, his central concern in 1984 was plainly the diminishing power of socialist utopias to carry political conviction. His own summary of his train of thought reads thus: 'Today the utopian vision of a labour-based society has lost its persuasive power – not only because the forces of production have lost their innocence or because the abolition of private means of production has obviously not been enough in itself to produce a society of self-governing labouring people. Above all, socialist utopias have lost their point of purchase in reality; the power of abstract intellectual labour to create new structures and shape society looks to be in almost total retreat'.[12] Thinking back to Horkheimer's characterization of critical theory, the conclusion seems unavoidable: on the Habermasian analysis, both the utopian notion of rational social conditions and the descriptive notion of the material boundary conditions necessary to make Utopia concretely realizable (viz. the development of productive forces) have lost their demonstrable basis in reality.

Let us note, first of all, that the collapse of the concrete vision of utopia, which Horkheimer's critical theory took its bearings from, does not necessarily rob critical theory of all concrete conceptions of reason. In his answer to Horkheimer's original essay, Marcuse had already pointed out that critical theory is *not* just a continuation of the Marxist critique of political economy; the bourgeois philosophical heritage is also something it can and should draw on: 'The idea that human beings are essentially rational, the idea that freedom nourishes that essential aspect of their existence, the idea that happiness is their most valued possession: they may be vague generalities, but their very generality gives them a certain progressive power now and in the future'.[13] In his response, Horkheimer seems to agree with Marcuse's view of the situation. Yet what is decisive is that Horkheimer's and Adorno's later critiques of reason, particularly Horkheimer's *Critique of Instrumental Reason* and the co-authored *Dialectic of Enlightenment*, underscore the increasing ambivalence of concrete conceptions of reason. As is well known, in the *Dialectic of Enlightenment* utopian conceptions of reason are shown to gradually unravel as reason reveals itself as a strategy of domination. Similarly, in the *Critique of Instrumental Reason*, Horkheimer points to a degeneration of reason that comes about precisely as reason is *realized* in the form of *Rationalität*, i.e. as more and more dimensions of social life find themselves subject to rationalization.

Actually, there is no need to show in detail that Horkheimer and Adorno's work contains the seeds of a veritable demolition of concrete conceptions of reason, it becomes all too obvious from the fact that all subsequent attempts to reconstruct a general picture of reason have taken the form of reconstructions of *subjective* reason. Reason, in short, has been reinterpreted either as a faculty belonging to individual human subjects or as an intersubjective practice. The intellectual work involved has been vast and it would be unfair to begrudge the protagonists of this Herculean post-metaphysical labour the pride with which they look back on what has been achieved. The era immediately preceding their heroic efforts presented a series of setbacks to the notion of rationality that no one could deny were of a *very* high order – not just the horror of the astute madman Nietzsche pictured late in the nineteenth century as the harbinger of a terrifying loss of normative orientation,[14] but also the early twentieth-century's experience of 'civilized barbarism', and the later century's thorough subordination of cultural and individual life to the maxims of rational economy and functional efficiency.

Considered against the background of these large-scale assaults on reason, the philosophical work of the period since 1950 has indeed done much to reinvigorate our trust in reason. The notion of *communicative reason* set out by Habermas, Apel and the New Frankfurt School has not merely helped a host of implicitly operative rules of discursive engagement to gain explicit acknowledgement in public discussion; through the concept of *communicative competence*, the heirs of Horkheimer and Adorno have had a theoretical and practical influence on education and have ultimately helped to endow democratic debate with something of the dignity of ethical discussion. Wolfgang Welsch's concept of *transversal reason* has played its part in counteracting the collapse of the public sphere into a dysfunctional plurality of discourses and subcultures.[15] And Ulrich Pothast's book, *Living Rationality* (*Lebendige Vernünftigkeit*),[16] has managed to imbue ordinary rational activity with a new seriousness by connecting everyday reflection and argument and the everyday language of sensation with the notion of personal life. However, none of these valiant efforts should mislead us into thinking that we are dealing with anything more than reason in its subjective guise. As Herbert Schnädelbach points out: 'What remains [after such intensive efforts at rational re-interpretation] is reason as a human capacity for being reasonable – in which case it would be more accurate to speak of reasonableness rather than reason or rationality . . . '.[17] Likewise, Wolfgang Welsch is surely quite right to call the various recently reconstructed forms of subjective reason *schwache Vernunft* – weak reason[18] – for while they may indeed help human beings to understand one another better, or orient themselves better in the world, they give little support to any sort of claim to change the world or criticize existing conditions. In particular, subjective reason in its latter-day guise would seem to shed nothing by way of critical light on the disastrous side of latter-day technification or on the socially constituted world of contemporary nature. *Schwache Vernunft* may be helpful for formulating sensible maxims of action, or for assisting in deliberation in normal circumstances between reasonable parties. But that is about the limit of its use. (A characteristic example of the weakness of subjective reason comes in the form of Apel's *ad hoc* attempts to uncover a rational conceptual apparatus, or what he calls bridging principles, to make possible the sorts of normative claims that are obviously simply *presupposed* by any discourse free of relations of domination. However, the *reductio ad absurdum* of the whole subjectivist programme would surely have to be Robert B. Brandom's *Making It Explicit*.[19] Brandom's conception of concepts and ideas would appear to deprive them of all power to create comprehensive evaluative standpoints or regulative ideals of any kind. Here, concepts and ideas have become little more than interconnecting points in an inferential web. Brandom's book lacks all notion of rational criticism.)

My argument would be that earlier critical theory looked to *substantive* or *objective* reason as a basis for criticism and hence approached the issue of the social use of reason from an entirely different angle. To see exactly how it did so, let us be more specific about the differences between these two forms of reason.

Substantive reason, we can say, is based on particular concrete ideas such as God, freedom and immortality in the Western metaphysical tradition, or the ideas of the soul, the world and God in Kant's transcendental reinterpretation of that tradition. And just as Kant thinks of reason as an endowment of human beings that orients them naturally towards transcendental ideas, so Marcuse sees reason as the essential ground of human life and Horkheimer writes that 'the goal of a rational society . . . is something every human being is indeed predisposed towards'.

Objective reason, on the other hand, is a form of reason whose origins go back as far as Anaxagoras and the Greek philosophical tradition. This, we might say, is reason as it is to be found *in reality itself*, either in nature or embodied in society. It is in this sense that Hegel could speak of a phenomenology of spirit – by spirit he means something like the unfolding of reason within history; in a similar sense, Marx looked forward to reason winning through in the history of class conflict. Horkheimer too refers explicitly to objective reason in setting out his concept of critical theory, situating the latter within an overarching objectively rational process, viz. the historical process of political emancipation, of which his intellectual project is said to represent one moment. For critical theory, on Horkheimer's conception, reason is the intellectual challenge that we have to set ourselves. Far from being a capacity of individual subjects, reason in its objective form is the touchstone of *necessary* ideas in light of which it becomes possible to critique existing social conditions and locate critical theory's very own theoretical practice in an objectively rational historical process.

Towards a critical theory of technology and nature

It strikes me that we are yet to really take up the challenge that is at the heart of the critical theoretic notion of reason: as far as nature and technology are concerned, we have a large body of criticism, but as yet no critical theory. To be sure, there is a long-standing critical tradition that has cast a consistently sceptical glance at technological developments; in the form of so-called *technology assessment*, critically oriented research into the birth of new technology has even established itself as a scientific discipline in recent decades. The concept of technification that took shape within the Darmstadt research circle, which provided a point of departure for the current book, seemed to give rise to a yet more radical form of technological critique, concerned not just with the tangible effects of technological development, but with the way that technological development has taken the form of a technification of social life and of human existence as a whole. Both technology assessment and the findings of the Darmstadt group were certainly critical in intent. Yet technology assessment remains tied to the idea of weighing the measurable costs of technological development against its measurable benefits. The Darmstadt group's concept of technification, on the other hand, remained somewhat nostalgic; the critical drift of the group's results came in the main from holding up the mirror of *pretechnological* conditions to current developments – so much must be said by way of self-criticism. Most tellingly, both technology assessment and the findings of the Darmstadt group lack both a conception of rational technology and a notion of what rational social conditions would look like in the context of a technological civilization.

However, the other side of the problem, the increasingly social constitution of the natural world, has also attracted a wide variety of criticism. The starting point of many critics is essentially ecology, which provides a tool for diagnosing the erosion, the disturbance of natural cycles, the overexploitation of groundwater, the contamination of ecosystems and the loss of biodiversity that result from particular social uses of the natural world. The weakness of this form of critique is evident above all in the environmental politics that it inspires, a politics of environmental conservation whose stance vis-à-vis society's use of nature is essentially restrictive. What is missing here is precisely a critical theory of nature, i.e. a theory of nature that treats human beings as an *essential* causal factor and can say in detail what rational social dealings with the natural world would amount to.

Similar problems and possibilities confront critical theory when we turn from external nature to the nature that we ourselves are. Here too we find well-established critical traditions, working with concepts like alienation, reification, technification and instrumentalization and the notion of the body's constitution through various disciplinary techniques. All provide rich material for biopolitics. None, however, has yet provided a solid basis for a positive concept of the body that is capable of orienting or justifying critical activity. Again, a fully elaborated critical theory – this time a critical theory of human embodiment – is lacking,[20] and with it a conceptual model of what it would mean to relate rationally to one's own body and the bodies of others.

Yet a critical *theory* of technology and nature is not all that needs to be developed, there is also, on the whole, a lack of practical environmental, technological and bodily *activity* that takes its explicit bearings from the notion of rational conditions. In general, the alternative practices that do exist – alternative technologies, alternative agricultural practices, alternative therapies for body and soul – appear to be developing largely in isolation. As in the case of the older versions of critical theory itself, today's suite of alternative technologies, farming methods, therapies and lifestyles remains largely disconnected from the sort of objective trends that could further the institution of rational norms.

Nonetheless, an encouraging start is being made. In many fields, the sorts of paired concepts that would pave the way for a co-ordinated description and a critique of existing conditions are available in theory. The most obvious example is the double concept of the *physical* and the *lived* body, which has helped to shape thought and practice in relation to the nature that we ourselves are. The point of the distinction should be clear: what is given objectively as human beings' physical nature can always be specified in terms of human *self*-experience; the body here is not just a physical given, but something posed as a human problem – not just a fact of nature, but a project of self-reflective human beings.[21] As far as the nature that we ourselves are *not* is concerned, i.e. external nature, research under the rubric of social natural science has put forward the double concept of *ecosystems* and *eco-structures* as a possible basis for sophisticated environmental analysis, critique and action.[22] The distinction is the same as that between lived and physical bodies: a part of nature that is taken in the context of the natural sciences as a straightforward object of enquiry is seen simultaneously from the social point of view of its limits and its normative value; natural reproduction becomes a product of human labour, not just the result of natural processes. Lastly, we might mention the double concept of technology as a *dispositif* that came together within the Darmstadt circle and opened up a new line of sight on the processes of technification. Technology, on this picture, becomes the ground of the possibility of a range of human activities, but at the same time a dimension of human activity that creates fixed structures; as well as freeing up time, minds and existing social institutions, technification brings with it a channelling and narrowing of human life as well. Yet as impressive as the work on these three sets of conceptual pairs may be, the question still arises whether they lack a *deeper rationale* in the absence of some sort of anticipatory picture of rational conditions. In the absence of shared presuppositions about what constitutes rational conditions, the fields of tension that each of the conceptual pairings create seem likely to open up few clear possibilities for action.

It is undeniable that a raft of solutions to the problem of rational conditions has already been proposed. There is certainly no undersupply of watchwords and slogans. Let us tally them up, for the time being without attempting to evaluate the sorts of claims that they are used to make. In the field of technology, we have the concepts of technological symbiosis [*Allianztechnk*], adaptive technology [*angepasste Technik*] and eco-technology. As far as external nature goes, we have seen suggestions for formal treaties with the natural world, or informal declarations of peace with nature, in addition to demands for greater sustainability and the idea of a humanization of nature. Lastly, in the realm of the nature that we ourselves are, the concept of human dignity has established a foothold, as indeed have the quite heterogeneous notions of holism and the deliberate technification of human bodily life (viz. through the development of human/robotic hybrids). Neither the potential nor the problems implicit in these concepts have been given much thought, let alone elaborated into a critical theory that is capable of providing intellectual foundations for critical arguments or points of reference for political action. The sort of conceptual labour along these lines that would be possible, and worthwhile, can be indicated briefly in the form of three sets of questions posed first of technology, second, of external nature and third, of the nature we ourselves are:

Can the notion of technological symbiosis (*Allianztechnik*) be extrapolated in concrete terms to the point where technology ceases to sustain the illusory project of mastering nature? Can *Allianztechnik* make possible some sort of interplay between human labour and creativity on the one hand and the laws and spontaneous activity of nature on the other?

How can we shape nature respectfully according to our own purposes in a way that enables human beings to lead lives of human dignity? How do we systematically take heed of the fact that the reproduction of society demands the reproduction of the natural world as the basis of all human life and all human society?

Can the notion of the nature that we ourselves are be integrated into our image of ourselves as human beings? Can it function as the touchstone of a notion of human dignity in the same way that notions of the soul and the person did in the not so distant past?

Computers in schools: Critical reflections on culture, technology and education

The concept of knowledge society

The concept of a knowledge society has had widespread currency in Germany since being imported from research circles in the late 1990s by the German Federal Ministry of Education and Research. Naturally enough, the researchers in question responded with gratitude to this rare burst of political enthusiasm for their work, though the adoption of a single conceptual term was obviously no guarantee that its critical potential[23] would be recognized. If anything, the opposite occurred: the notion of a knowledge society served largely to strengthen the ministry's overall hand within the Kohl government, lending a kind of intellectual legitimacy to particular departmental programmes, particularly a 1996 initiative known as *Schools Online*. In political practice, the point of labelling contemporary society a knowledge society seems, in retrospect, to have been to prepare the ground for a restructuring process within education whose central premise was that IT skills, computer literacy and media awareness would be the pivotal forms of human capital for future society.

If *that* had been their only point, the ministry's political agenda setters might have chosen any of a number of theoretical tropes developed within the sociology of science since the 1960s. The choice of *"knowledge society"* might almost have followed from a process of elimination: the notion that science was being transformed into a potent *force of production* was politically suspect because of its Marxist origins, as was the notion of a *technoscientific revolution*. The concept of *post-industrial society* on the other hand was obviously never going to have much appeal within an educational agenda aiming to encourage maximal participation by industry, while the notions of *information society* and *scientific society* might have been difficult to present as anything but a *terminus ad quem* to teachers working under secure conditions within the public service and still much influenced by the humanistic ideals of Goethe and Humboldt. The notion of knowledge society had no such drawbacks. In retrospect, computers could hardly have made their triumphal entry into the classroom under a brighter standard.

Because of the resoundingly positive connotations of the concept of knowledge, the notion of a knowledge society quickly took on a heavy freight of ideological meaning. Politically, the term came to be used in a purely descriptive sense, viz. as a catchall for a certain gigantic social trend – the tendentially ever-increasing importance of knowledge for the growth of general social prosperity. As an antidote to this kind of hazy political usage, it is worth recalling that the theory of knowledge society had been intended quite differently by its originators, viz. as a *critical social theory*. The term 'knowledge' came to have pride of theoretical place *in contradistinction to* terms like information and science in order to bring out three essential points: that *alternative* forms of knowledge exist; that forms of knowledge inevitably come to stand in *hierarchical* relation to one another; and that such hierarchies have *social consequences* for the *social carriers* of the various knowledge forms.[24] Critically speaking, the original theory had several aims, not just to provide a conceptual background against which to *evaluate* as well as describe the scientification of human social relationships (for instance, individuals' growing dependence on scientific experts), but also to draw attention to the way that knowledge has grown in importance, both as a basis of social power and as a determinant of individuals' life chances, at the very time when other primary factors, such as birth and capital, have become (relatively) less significant.

To reassert and build on the critical potential of the theory of knowledge society, we need, above all, to do away with the truncated singular terms that dominate so much of the ideologically coloured discourse of education policy and start working with the sorts of dialectical conceptual pairs that we are already familiar with from critical theory. Just as critical theories of political economy work with the twin notions of *value* and *price*, and critical theories of human nature draw critical distinctions between the *lived body* and the *physical body*, so too the theory of a knowledge society operates with the twin notions of *knowledge* and *information*.

Bringing the latter two concepts into connection with one another while, at the same time, allowing them to denote a clear conceptual difference is of the essence.

Tapping the critical potential of the theory of knowledge society involves much more than anchoring a discussion of the social function of knowledge/information in *philosophical* notions of knowledge, at least as they come down to us from the history of Western philosophy, for since Plato philosophy has repeatedly[25] tended to establish or reinforce fixed hierarchies between forms of knowledge or has sought to discredit alternatives to scientific knowledge, thus condemning non-formal and non-systematic modes of knowing human experience to historical obscurity. Nor will it do to mint theoretically sophisticated versions of everyday linguistic notions of information[26] (what it means to 'inform oneself about an issue' is an insufficient base for theory), nor to simply recur to the technical usage of the term 'information' in a specialist field like Information Theory. Of course, in a way, there is no getting around the conceptual history of our key terms, and their specialist uses are indeed something that we have to take as a point of reference. What is most important, however, is to bring knowledge and information conceptually into relation with one another and to do so in a way that casts critical light on societies in which having bodies of objectified knowledge at one's command has become a key problematic.[27]

My suggestion here would be to define knowledge as a *mode of engagement*, not just with scientifically defined things or states of affair, but with a broader web of social relations too. It is through knowledge, we might say, that things and states of affair are *given* to us. Our concept of knowledge, on this picture, becomes a relational concept and implies, on the one hand, a reference to reality and, on the other hand, a reference to a knowing subject, *a person*. What is decisive here for the conceptual pair knowledge/information is that knowledge can be *objectified*. To put the point trivially, it can be written down. Knowledge in written form is, however, knowledge that has been transformed into *data* of one sort or another, objectified knowledge, and such objectified knowledge is something that we can usefully call *information*. If objects or states of affair are what are given to us through knowledge, then through information (objectified knowledge) what is given to us becomes available as a symbolic representation. Here, one can already see that in order to transform information *back into* knowledge, information needs to become the basis of a relation to reality established by a human subject, i.e. a person. In essence, the move from information to knowledge occurs when a reference to reality – and that means both a reference to human beings (to people) and a reference to objects and states of affair in excess of the system of symbolic representation – is activated or reactivated.

Now the all-important feature of social life in a knowledge society is the existence of a type of knowledge – we could call it *second-order knowledge* – that isn't based on immediate interaction with objects of knowledge or states of affair, but on interaction with *already objectified knowledge*. If this objectified knowledge (information) is at all meaningful, if indeed it can come to have greater importance than first-order knowledge, that is because it can normally be assumed that second-order knowledge has objective reference and because social reality is something quite different from reality *tout court*,[28] viz. a web of representations or symbolic forms, rather than a set of well-defined objects and states of affair. For the scientist, there is usually no need to establish a solid reference to objective reality at absolutely every point of his work. Practising scientists themselves can rely on the fact that such reference has already been established by other scientists. In short, they can refer without further ado to already established results and laws. Scientific knowledge itself is thus largely a form of engagement with already objectified knowledge. Something similar is true of *cultural* knowledge because cultural objects exist in symbolic forms as well; as a rule, they are already given in the form of data – that is, in texts, notation and, more recently, in digitalized media.

Because this type of second-order knowledge has come to prevail over immediate first-order knowledge and the sort of interaction with things and states of affair that first-order knowledge demands, knowledge in today's society has, on the whole, come to mean the appropriation of information. The all-important question that our society finds itself faced with is thus which skills its members need to acquire in order to successfully bring given bodies of information into meaningful reference to people, objects and the world at large. The brute fact that information is technically accessible emphatically does *not* mean that individuals will possess more knowledge, nor does it mean that our society will, on average, be a more knowledgeable society. Quite the opposite: the growth of widely accessible information can lead to a decline in the overall level of our knowledge because access to knowledge requires a set of highly developed skills in appropriating knowledge. To use a classical, and relatively simple, example: to make use of the information contained in a library, one obviously first needs to be able to read. The wonders of a library, however, will only open up when one has something more on top of that: a systematic way of accessing the library's contents.[29] Perhaps a better example would be the famous Pythagorean theorem that the square of the hypotenuse of a right-angled triangle is equal to the sum of the square of the other two sides. To intellectually appropriate the content of the theorem in this case implies both understanding it in a fundamental mathematical sense (in short, being able to *prove* why it holds) as well as grasping how it can be used to generate higher-order mathematical constructs and applied in practical situations. Speaking generally, we can say that fully appropriating information (successfully transforming information into knowledge) involves a complex ability to seek out information and to evaluate and contextualize it, to form a picture of the way it came to exist as part of a body of information, to interpret it, to relate it to other information and then, finally, to use it. The question we want to ask about the use of technology within the contemporary education system is thus essentially: in what way does the technification of education assist, and in what way does it hamper, the transformation of information into knowledge?

A fourth form of cultural technology?

The use of computers throughout society is reason enough for theorists to rearticulate and further develop the theory of knowledge society, for while other periods of human history have indeed privileged knowledge in similar ways – ancient Egypt is probably the best-known historical society to have made knowledge the basis of power and the key to the distribution of life chances – the need to make an in-principle distinction between knowledge and information has its origins in much more recent historical changes. Essentially, the importance of the distinction in the world of contemporary modernity lies in the widespread accessibility of enormous quantities of information, the technical accessibility of what we have called objectified knowledge. It is precisely this technical accessibility of information that creates a problematic differential between our command of information and our ability to transform that same information into knowledge. Prior to the technification of our means of processing information, such a differential could hardly be said to exist. Humankind's historical predicament down to very recent times, in short, could be captured within the terms of a relatively simple equation: someone who could read a text could also understand it – in more or less sophisticated terms, of course.

Books were obviously the classical medium in which knowledge was objectified, and from the point of view of world-cultural history we can see them purely as a means of storing information. Reading was the ability to appropriate such information, where by reading we of course do not mean the vocalization of a physical text, but the process of responding to written symbols in a way that endows the text with a *meaning*. Reading simultaneously involves *interpretation* in this emphatic sense.

Now the general situation in the contemporary world looks roughly like this: computers and the internet, together with their associated data storage systems, represent an increasingly universal medium for storing objectified knowledge (i.e. information). And the capacity to appropriate such information is something we can generally call computer or internet literacy, as long as we bear in mind the risk that we thereby run of obscuring the difference between a *technical command* over information and the *meaningful conversion of information into knowledge*.

The risk is far from academic, for this is precisely the danger incurred by an education politics that sees computers as the central challenge facing the contemporary school system. Education is taken to mean the transmission of knowledge, and since knowledge is increasingly available throughout the world in objectified electronic form via computers and the internet, computer literacy is increasingly seen as the key skill set that needs to be transmitted to younger generations. Computer literacy, we might say, is thought of as a fourth form of cultural technology, next to reading, writing and arithmetic – quite understandably, given that all sorts of social transactions (banking, shopping, etc.) already take place digitally, and given, moreover, that computer competence and internet skills are a fundamental prerequisite in an increasing number of sectors of the workforce. It seems as if the time is fast approaching when participating in social life will be all but impossible for anyone who has failed to attain fluency in the language of the digitalized world, just as, not so long ago, taking part fully in social life was all but impossible without an ability to read, write and do basic maths. It is in this sense that we can think of computer and internet literacy as a fourth form of cultural technology. What is becoming clear in recent years is that the new cultural/technical skill set could come to displace or even replace the three classic cultural/technical skill sets. It is already the case today that many people's basic maths skills amount to nothing more than an ability to use a calculator. Should a similar situation develop as far as reading and writing goes, the result would be a nightmare. And does it depend ultimately on anything more than the ingenuity of software developers whether computer-aided writing becomes possible in years to come? Of one thing we can be sure: that the *temptation* to use such – at this stage notional – technical substitutes will be conflated all too easily in the real world with the *necessity* of using them.

There is no denying that if computer skills and internet skills do indeed amount to a fourth form of cultural technology, then the task of transmitting and fostering these skills is one of the duties of education. However, a sense of the relative importance of the classical cultural/technical forms and the cultural/technical newcomer needs to be established. Above all, the three classical cultural technologies must be seen as forming an absolutely indispensible *presupposition* of competent computer and internet use. Someone who is incapable of basic maths is going to be incapable of using electronic calculators, just as someone who is incapable of reading and writing is going to be incapable of using word-processing devices in all but the most rudimentary sense. One can certainly imagine different possible scenarios, and the loss of basic literacy and numeracy, and an ensuing cultural meltdown, should by no means be taken as a preordained outcome of increasing computer use. However, given the present direction of education policy and the present impetus of technological development, the pressure to replace literacy programmes with computer courses appears to be growing throughout the world. Modernization pushed in this direction, we would have to say, is a highly ambivalent phenomenon; in the worst of all possible worlds, it would be capable of reducing human beings to the level of the apes that primatologists communicate with by teaching them rudimentary keyboard skills.

What do people need to learn in order to make competent use of computers and the internet? One thing at least is clear – that the current pressure to introduce computers into contemporary education is quite different from the push to introduce so-called language laboratories into European schools from the 1950s onwards. That pedagogical experiment was an instructive one; practically all of the expectations associated with language laboratories proved to be illusory. In particular, the prospect of reducing the amount of face-to-face teaching involved in the learning process proved completely illusory. Despite superficial similarities though, there is no comparison with the current wave of computerization that is taking place in schools, for computers today are not simply presented as a teaching *tool* or as the latest in a series of devices for faster, easier and more enjoyable learning (on this score, they probably live up to their promises), but as a separate subject area of a special kind. Computer skills are not just taught so that they can be applied in other areas of the school curriculum, they are taught for their own sake. The question that immediately arises though is this: what sort of skills *are* computer skills? What is actually being taught here?

The most obvious answer – a short-sighted one – is quite simply: information theory. The argument runs as follows: if computers are essentially information-processing systems, information theory is the appropriate subject in which students can acquire the necessary skills in dealing with the relevant systems. One can see how this fits with information theorists' understanding of themselves and their own importance. Wilfried Brauer, for example, formulates the idea as follows: 'Information theory is a science that deals systematically with the processing of information, particularly automatic processing techniques that make use of digital computers. Digital computers and the technology of associated data-processing systems are the two fundamental pillars of data-processing'.[30] This is information theory at its most uncritical – and at its most influential, for there are many who would like to see precisely this sort of thing introduced into schools in order to meet the needs of the knowledge society of the future. In this vein, one humanities-based German high school went as far as offering information theory as an alternative to ancient Greek and French.

What becomes clear in the work of Brauer and his colleagues is that a truly *critical* brand of information theory is remote from mainstream work in the field. Like the critical theory of knowledge society, a critical theory of information, above all, would have to conceptually preserve the difference between knowledge and information. Here as elsewhere, theoretical sophistication begins with a step towards critical self-reflection. The prime focus of a critical theory of information would be, first, the transformation of knowledge into information and, second, the appropriation of information and its reconfiguration as knowledge as part of the lived experience of human beings; critically speaking, the factual question of *how* information is mechanically processed is of secondary importance. It is only such a critical theory that could inculcate computer literacy in the deeper sense of the word, viz. the dual ability to evaluate the *weight* and *meaning* of information presented in technological formats, and to modulate information for the purposes of knowing human subjects. In contrast to Brauer's strictures, the work of Alfred Lothar Luft seems to make a promising theoretical start. Luft describes information theory as a technological science whose main themes are 'the representation of knowledge in the form of data and the reduction of the activities of the human mind to algorithms and mechanically simulated processes of various kinds'.[31]

Short-sighted education policy

One of the largest projects to be initiated in recent decades by the German Federal Ministry for Education, Research, Science and Technology, a co-operative venture in association with German Telecom called *Schools Online* was launched in 1995. The programme aimed initially to connect 10,000 German schools to the internet. By January 2000, 16,500 of a total of 40,000 schools were to have internet access. By the close of 2001, every school in the country was to have been brought up to speed.[32] According to a ministerial press release of 16 October 2006, the latter goal was finally reached in the tenth year of operations. By this stage, 20,000 German schools, or roughly half the schools in the country, were equipped with a broadband connection to the World Wide Web.

Yet despite the overwhelming quantitative success of *Schools Online*, the programme seemed to have little by way of pedagogical rationale. A magazine as favourable to high-tech education policy as *Computers and Teaching* was among a number to review *Schools Online* in more than mildly sceptical terms: 'Real evidence that the use of new media leads to fundamental improvement in teaching quality has not been forthcoming'.[33] Three years earlier, the same point had already been made in the United States, and in much more forceful terms: 'A 1997 survey of 6000 US teachers, computer co-ordinators and school librarians found that 87 per cent believe that internet usage by students in grades 3 to 12 does not help improve classroom performance'.[34] Although the pedagogical benefits of computer use in the classroom such as interactive learning and better communication skills are repeatedly showcased as part of public discussion, sober analysis tends to suggest that they are at best side effects, rather than the direct goals of any sort of well-considered educational methodology. The aim, insofar as there is one, is essentially 'to get children in a fit state to face the fully digitalized world of tomorrow', as one commentator put it.[35] In short, we seem to have got to the point of simply *assuming* that life in the developed societies of the future will be fundamentally shaped by computers – or, as we have just put it, that computer literacy will come to be a fourth form of cultural technology.

Hazy general assertions often displace all mention of goals in the various discussions of the role of technology within education policy – 'in the switch from an industrial society to an information society or a knowledge society we are seeing a paradigm shift of the whole field of education', as Renate Schultz-Zander puts it in glib macro-historical terms.[36] And yet this way of presenting current trends as assured future realities might, in a way, be as honest a way of broaching the issue of educational technification as can be expected; any more normative claim that the move from an industrial society to an information/knowledge society *demands* a fundamental reform of the education sector would surely overlook the fact that computerization within schools is itself *already an inalienable part* of the transition to an information/knowledge society. In other words, the 'digital future' is something that is being actively brought on by the introduction of computers into the classroom. As a basic technicist mindset is inculcated at an early age, the use of computers in every aspect of social life – from work to leisure to transport to communication – becomes something habitual, self-evident.

The psychological or intellectual development of students is *not*, in short, what is driving the fast-paced introduction of computers into schools. As we have seen, rapid computerization has been put into effect rather half-heartedly by teachers themselves. So what does the paradigm shift that is said to be taking place throughout education actually amount to in substantive terms? On closer inspection, the answer seems to be – primarily, the transformation of education into a *very* capital-intensive domain. To be sure, this is something one might have guessed from the involvement of German Telecom in the Kohl government's *Schools Online* programme. But it becomes still more obvious when one considers the vast sums of money that are involved in the computerization of the entire school system. According to a combined study by the Bertelsmann Foundation and the University of Bremen, the cost of providing computer access to all German students at all levels is in the order of €81 billion. Yearly maintenance and operating costs add a further €24 billion to the total.[37] Although the figures relate to the as yet unrealized goal of 'full computerization' (universal broadband access and a personal computer at the disposal of each individual student), the costs of what has actually been achieved are still remarkable. The city of Munich, for instance, allocated a total of €200 million for computer programmes in schools in the year 2000 alone,[38] while provisions for computerization in the south-west German state of Baden-Würtemberg were set to cost €1 billion just for the year 2001.[39]

The figures become all the more breathtaking when one considers the sort of spending that is *ruled out* by the drive to computerize the learning process. Thus, in Munich, the follow-on costs of computerization alone would have been enough to fund a total of 200 new teaching positions. Conversely, every 1,000 computers require an additional two positions for computer technicians.[40] And while the anxiety that the increasing use of computers in the classroom will make face-to-face teaching redundant seems largely unfounded, it is inevitable that huge technical outlays put enormous budgetary pressure on personnel costs, putting paid once and for all to the goal of reducing class sizes (a demand loudly voiced by teachers for decades). All in all, the education system is becoming a capital-intensive sector of society, essentially to suit the economic agenda of industry. Little wonder that peak industry bodies were quick to 'spontaneously promise their full support' for *Schools Online* in response to an official call from the Minister for Education and Research for 'an industry sponsorship deal second to none'.[41]

The problems associated with turning education into a capital-intensive sector can be illustrated by a dispute between state and local government that played itself out in Baden-Würtemberg in the early 2000s. Until the controversy broke out, the cost of public education had been shared between the two levels of government, with the state meeting the cost of staff (wages, etc.) and local councils covering 'material costs' (the construction and upkeep of buildings and the like). Understandably, councils resisted footing the bill for new computer equipment and ongoing technical maintenance, which the state government was pushing, for obvious reasons, to include under the heading of material costs.[42]

Here, it would seem, we confront the *actual* paradigm shift that is taking place in education in an acute, and acutely problematic, form; what that shift entails first and foremost is a radical new approach to public expenditure within the education sector, a radicalization of the demands made on government, potentially, as in Baden-Würtenberg, a radical redivision of financial responsibility, and certainly, as we have seen elsewhere, a radical change in the nature of corporate involvement in public schooling.

For years, the introduction of computers in schools has been promoted as the truly innovative aspect of education policy, yet the clear pedagogical conceptions that one would expect to underwrite such a shift of priorities and resources are almost wholly lacking. While many seem to imagine that computers are a useful teaching aid, ideally suited to increasing the efficiency and speed of learning across all subjects, in practice what we see is a push to teach computer and internet skills for their own sake. The first round of studies of computerization in schools indicates that the subject area where the new technology is put to genuinely practical use is, unsurprisingly, information theory.[43] In recent years, information theory has been systematically established as an independent field of study, and it has often been promoted in schools through extracurricular workshops. It first appeared on the syllabus towards the end of the 1970s; in the 1980s ITG (*Informationstechnische*

Grundbildung – Basic IT Studies) was introduced into junior high schools throughout Germany. Admittedly, putting the idea of system-wide ITG classes into institutional practice turned out to be impossible in the early years because of a shortage of qualified teaching staff. That situation began to change slowly in the 1990s;[44] however, what is lacking to this day is a substantial concept of what it actually means to teach or study Basic IT. Just as importantly, there seems to be a lack of clear thinking about *when* computer classes are appropriate, i.e. the age at which children are best immersed in the digital world. A typical example of German political leaders' dubious urge to keep up with the technical spirit of the times is to be found in the Bavarian State Government's primary school policy guidelines, quoted in the above-mentioned issue of *Computers and Teaching*: 'In the context of basic IT studies, interaction with computers is clearly indispensible' – even at the primary school level. The same article reports the creation of a competition whose entrants are encouraged to give as sparkling a demonstration as possible that 'computers can be put to creative and meaningful use in primary schools too'.[45] Other sections of the Bavarian white paper earnestly discuss the possibility of installing computers in kindergartens.

Rational education policy?

Given that we no longer adhere to ideals of rationality in the deeper metaphysical and political senses specified in our previous essay, it is certainly not easy to say without further ado what a rational education policy would amount to in the context of a fast-evolving knowledge society. But we can at least make some distinctions between what is rational in the sense of being reasonable and what is clearly not. Our argument here would be that an education policy is rational insofar as it is conceived not just with a view to the social realities and economic imperatives of the day, but also with a view to developing the all-round potential of younger generations.[46] A rational pedagogical model will certainly take on board that we live in a society in which computer and internet literacy has become a cultural/technological form in its own right. However, it will equally strive to relate the new form of literacy to the cultural/technological forms of old. Moreover, it will be highly mindful of the need to harmonize the teaching of basic skills with the age of the students. It is patently irrational, in any case, to make the school system subservient to a short-term social need; for instance, to rationalize compulsory computerization in schools with reference to the current high demand for IT specialists in various sectors of the economy. Likewise, it is rationally unjustifiable to drive the long-term process of adapting education to the dynamics of a knowledge society solely as it were from a hardware point of view – for in practice that means flooding schools with computers, which in turn means putting insupportable strains on education budgets and/or accepting the blandishments of commercially self-interested industry groups.

In defiance of the current uncritical consensus within policy-making circles, it is worth pointing out that computers have not fundamentally improved human life. They have not even brought about improvements to economic efficiency or productivity.[47] Life – to say it once again – is no better in the age of the computer than it once was, it is simply different. Looked at in broad terms, the primary effect of computerization is probably the growth of a dynamic new sector within the world economy, whose emergence, all things told, has done little more than compensate for the concurrent decline in other major economic sectors. School life within a technological civilization is clearly also going to be something different to what it once was; however, it is irrational to attempt to prepare human beings for the broader changes that are underway solely by exposing them to computers and the internet from infancy, or by mandated doses of information theory from primary school onwards.

It is important to remember that students today learn most of what they need to know about dealing with computers from each other, i.e. within their peer groups. As just about anyone can confirm from experience, teenagers are normally technically well ahead of older generations; they are normally far more technology savvy than their teachers. Moreover, the way that school-age students learn to use computers is essentially playful, innovative; in general, they adapt to each new technological innovation with extraordinary speed. Yet school computer classes can probably harness little of this sort of spontaneous adaptation to new technological forms because, by its very nature, teaching in schools has to be relatively structured and rule bound; what students learn in classroom situations is bound to be quickly outmoded for the same reason. Information theory, at least at its current stage of development, is also unsuitable as a school subject. Neither primary nor secondary schools are there to produce cohorts of junior information theorists capable of theoretically grasping what data-processing systems do in practice, viz. process data. Here we confront the prime blindspot that is manifest in the current lack of pedagogical forethought: it is not rational to expose a child to computers at a very early age, and especially not to their technical workings, because computers are, by their nature, devices that operate on analytic and algorithmic principles, while the skills that children develop in these formative early years are principally intuitive and imaginative – the plastic ability to grasp forms and shapes, *not* a proficiency in manipulating symbols and mathematical functions. Such abilities need to be carefully nourished before children are presented with personal computers and their associated technical paraphernalia.

The knowledge societies of today are characterized by the existence of an enormous capital stock of knowledge in the form of technically accessible information. However, they are also marked by the fact that the individual's life chances, including everything from social status to opportunities for self-fulfilment, are determined by his opportunities to appropriate the available stock of knowledge. In addition, contemporary knowledge societies are dependent, at the national as well as at an international level, on the accessibility of knowledge; as we have suggested, they tend to be structured hierarchically, with the individual and the nation's position within hierarchical power relations determined by the amount and quality of the knowledge at their command. Education policies that aim to prepare younger generations for this situation must do so by emphasizing two overlapping distinctions: between information and knowledge, and between having a *technical command* of information and its *meaningful personal assimilation* as part of *lived experience*.

This is certainly no plea for banishing computers from the classroom. It is, however, an argument for making rational use of the computer. However, the rational use of computers, we believe, presupposes the *strengthening* of classical literacy and numeracy. If computers are not simply to be used as elaborate calculating devices, students will need to be able to form sophisticated pictures of mathematical relations, they will need to understand mathematical operations and be able to prove mathematical theorems. Likewise, if computers are to become more than access points to the internet and the latter more than a tool for downloading and compiling miscellaneous textual snippets, students will need to exercise critical judgement in their use of sources; they will need to be able to interpret texts and bring ideas from isolated contexts into meaningful textual relation. In short, the competent use of computers and the internet presupposes the development of the classical cultural technologies, indeed it emphatically requires an *extension* of reading, writing and arithmetic.[48]

An education policy is, in this respect, no different to other areas of social life: rational social behaviour and rational social agency will be difficult to achieve without a degree of dispassionate assessment of the major trends of social life, and without some preparedness for resistance and non-conformity. If the job of educators and policy makers is to prepare students for life in a knowledge society, and if the best basis for doing so is open acknowledgement of the basic social facts of life within knowledge societies, then it is the opposite of rational policy to simply abandon students and the education system as a whole to prevailing social currents, for these currents point clearly in a questionable direction: first towards the transformation of the education sector into a capital-intensive, increasingly commercial operation, second, towards the technification of our dealings with knowledge and third, towards the dependence of individuals on the provision of ever more complicated forms of information technology.

Thinking anti-cyclically

What do we need to know in a knowledge society?

We live in a knowledge society and we have to cope with the profound challenges that that presents, getting our bearings within evolving social structures and preparing our children for life in the world that is coming into being. But what is a knowledge society? Let us review the answer we have given, both here and elsewhere.

My initial idea in introducing the term 'knowledge society' in 1985 in a book co-authored by the sociologist Nico Stehr[49] was roughly as follows: Stehr and I wanted to emphasize that political power and individual life chances, both within the developed world and between developed and developing nations, were becoming increasingly relative to the level of one's participation in knowledge creation and knowledge use. The concept of a knowledge society was intended in an emphatically critical sense. Stehr and I were seeking to create an awareness of the variety of forms of knowledge and of the fact that different forms of knowledge have different social functions and different social carriers.

A knowledge society, I maintained and still maintain, is something quite different from an information society, which was the main sociological *typos* that other theorists had proposed to explain how data processing was fast becoming the biggest sector of economic life, with a rapidly expanding role in all parts of the economy. The arguments of the advocates of an 'information society' seemed singularly unable to account for the wider social meaning of the use of data and data-processing systems; because data processing – they suggested – makes up the greatest part of total social output in information societies, so the essential skill that human beings need to acquire in order to negotiate the challenges of life in such societies is an ability to deal with information, whether that means searching for information, processing information, storing information, surfing the net, creating software or finding applications for software. Our counter to this line of argument in the previous essay was that such a conception of information and its uses is a poor guide to what schools are there to communicate to their students. Children learn about computers and other technological devices in a playful way and they do so primarily within their peer groups. This is, of course, not to argue that computer learning belongs wholly outside the classroom; here and there, schools can of course pick up on the skills that students acquire among their peers, and they can and should encourage students to exchange skills in group work, making use of new technology as a tool wherever it is relevant, though not for its own sake. And yet it is nonsensical to extend the use of technology in the classroom beyond those basic limits, that is, to make internet skills a separate school subject or to replace language classes with information theory, especially given that the very concept of information theory has as yet little clear meaning at the level of theory.

The critical theory of information remains in its infancy. And so under the pressure of immediate social demands – in the rush to prepare enough technical specialists for the so-called 'media boom' – policy makers, theorists and commentators have limited themselves to the purely technical side of the question of information. What that generally means is a narrow focus on data processing. As a result, information theory in its present form is without serious theoretical foundations. A minimum level of reflection about what the application of information theory itself means is yet to be attained.

What is information in the first place? We have already suggested a broad answer: information is objectified knowledge, knowledge objectified in the form of data. So much is clear. But what we need to know next is – what has to happen if a phenomenon or a set of phenomena is to be understood in the form of data? What sort of preconditioning must phenomena undergo if the form of knowledge known as data processing is to be applied to them? For example, what forms must social organization take if we are going to understand society in terms of data? The answer to such questions would be the very least we need if we are to get our bearings within a knowledge society. Information, as Carl Friedrich von Weizsäcker once said, is what can be understood; in the language of our argument, information is data with a social meaning. In fact, that definition points to the second major thing that we need to know if we are to understand the knowledge society taking shape around us – we need to understand in general terms what the process of giving meaning to data involves. As things stand, we could hardly be further from addressing either of these two major problems. Today's information theory limits itself to the question of how data are produced from other data.

Now it is clear that dealing with data is essential to the way that our society functions in the present day and age. And yet what we actually have to know in order to deal with data competently is something that is, in a sense, prior to information theory and, in a sense, beyond it: prior to the theory, we have the plain abstract fact of data generation (a set of categorial distinctions and measurable quantities), while after the theory we have the practical use of intelligence involved in making sense of data. Philosophers might say we have, on the one hand, an epistemological problem (data generation), and on the other hand, a problem of hermeneutics (data interpretation). However, although both problems are highly relevant to a proper theory of information, neither has anything to do with the issues that face our educators: in the classroom the key question is literacy – the basic ability to read and write, or in other words the ability to transform the world and one's experience of it into *texts* and the related process of making sense of already existing texts. In short, the real issue facing educators today is the classical skill sets that the profusion of data and the ease with which data can be obtained make even more fundamental than ever. And yet today's policy makers would seem to be rapidly beating a path in the opposite direction. Information theory, as we have noted, is being offered in place of French and Greek, extra IT technicians are being recruited instead of teachers.

Forms of knowledge: Science and others

To call our contemporary knowledge society an *information society* is to invite a crude reductionist misunderstanding. However, it is equally one dimensional to label it a *scientific society*. Science and its associated repertoire of techniques constitute one *form of knowledge* among many. Living competently in a knowledge society means keeping the differences between the various forms of knowledge intact and knowing what the different functions of the different forms of knowledge amount to. It is in this way and no other that human beings can acquire and use the knowledge that is necessary to cope with different fields of experience.

A typical example that I have discussed elsewhere from an historical point of view is the scientification of the process of childbirth,[50] or, to put it differently, the growth of obstetrics at the expense of midwifery. An important form of knowledge appears, tendentially, to be vanishing – the quasi-traditional knowledge of midwives that was based on first-hand experience, anchored in an immediate lifeworld and mindful of the social and biographical circumstances of each particular birth. Since at least the early twentieth century, a far-reaching technification of the whole field of childbirth has reduced midwives to the status of technical assistants to doctors, professionally qualified assistants to be sure, but with nothing like the personal body of knowledge or the social role of the midwives of old.

However, let us turn to an example that is more relevant to education. Put bluntly, we can say that contemporary high schools teach biology, chemistry and physics as if these subjects were the sole form of knowledge of the natural world. All other forms of knowledge of nature are disparaged as non-knowledge because they belong outside the enterprise of natural science. More than that, the particularity of natural science and its social role is something that students are left fundamentally in the dark about. As a countermeasure, one recommendation would be to teach Goethe's *Theory of Colour* alongside the physicalist optics that derives from Newton; students could then at least be reminded that modern physics is actually one among a number of possible understandings of physical nature, viz. an interpretation that aims at technical mastery of natural conditions. Against the objective, instrumentalizing approach to nature that is at the heart of the physical sciences, schools could contrast a knowledge of nature acquired via direct sensory experience that allows students to systematically orient themselves vis-à-vis nature *as it is for us* rather than nature as it is *in itself*. Similar recommendations would be in order for other subject areas. With first-hand experience of physical nature rapidly disappearing from average everyday life within a technological civilization, and with biology increasingly reduced within science itself to biochemistry, the reconstruction of proper school gardens would be a small step in the right direction.

The internet and the body

In an age in which the socialization of younger generations takes place in a highly technified environment, one of schools' main priorities must be to offer a degree of resistance to the main forces shaping individual experience. Well before entering the workforce, children and teenagers come to know the world via technical networks of communication and transport, through the products of the entertainment industry, and through the host of devices that form part of everyday household life, computers foremost among them. In response, schools face the task of showing that the overlapping forms of technological life – the gadgets, networks and systems – are not the whole of life. The experience of a non-technified world, which can no longer be taken for granted, is in short something that education should actively seek to supply.

Schools must at last recall that their task is not just training children to fulfil future social functions or to realize their potential as economic producers, but to educate human beings. If the basic question of education in our time is what younger generations need to be taught in order to prepare them for life in a knowledge society, then the answer must speak essentially to the fact that they will want to lead meaningfully human lives; the aim cannot simply be to provide them with a set of skills that will enable them to achieve at a certain level according to conventional standards of social or economic success. If schools now believe that they need to teach youngsters how to exert themselves to good effect on the net, then they are neglecting their basic duties in one essential respect at least. Younger generations already know how to exert themselves to good effect on the net. But what that means is that they run the risk of overexerting themselves on the net. In short, they risk disregarding other sources of the self and other sides of life, above all the life of the body. The internet generation may well spend less time watching television than the so-called baby boomers, but all told they probably spend more time in front of monitors, given their overall use of computers as well as television.[51] The likely result is a neglect of the body and a generalized withering of the capacity for sensory experience, if not the sort of progressive hyper-cerebration prophesied by Gottfried Benn.

Here, one could fall back on the hopes of virtual reality (VR) aficionados who see the potential for a rediscovery, or even an extension, of physical sensibility in

VR glasses and gloves (perhaps soon some kind of VR jacket, theoretically perhaps even a sort of second VR skin). However, technological optimism seems to have gone conceptually off the rails here, for surely any notional extension of sensory experience would have to be based on our existing five senses; it could do no more than translate what is outside the range of the senses into the language of existing sensory experience. A good example is the visualization of phenomena by means of technical reconstruction – probably best known in the field of sonography, which transforms ultrasound data into a stream of images. In terms of new sensations made available by VR gloves, it seems reasonable to point out that what technological gadgetry can transmit is distinctly relative to the specifications of technological models. Using such mechanical sensory stimuli, in other words, limits the sensory experience of the human user to the possibilities inherent in the mechanism involved, a situation that promises not to extend the user's sensory range, but to restrict him to a particular pattern of sensory experience, and to drill him in responding in its terms. (An historical example of technification of this sort is the modern dominance of sharp focused vision as a paradigm of human vision, in Europe a consequence of the extensive use of optical devices from the seventeenth century onwards. And, indeed, something of the same effect can be seen in the present-day use of voice-recognition technology; our contemporaries, or some of them at least, dutifully rehearse pronouncing their vowels and consonants in a way that fits the requirements of mechanized speech.)

In the face of these sorts of impoverishing trends, what contemporary education urgently needs to prioritize is a quite basic ability to see, hear and feel. That may sound like a quizzical recommendation. However, what I have in mind is nothing particularly mysterious. A good start would be to expand the imaginative repertoire of school art programmes with the conscious goal of broadening children's capacity for bodily experience, rather than supplementing their stock of knowledge, e.g. of art history. Physical exercise needs to be added into the mix – preferably a form of PE that means something more than just sport and has a higher goal than creating a mass basis for national Olympic success. To get a picture of the sort of problems that present and future generations of net users face, it is worth recalling a little history. In the eighteenth century, the first of what we might call the great diseases of modern civilization, hypochondria, emerged as a result of sedentary patterns of life; weavers and university professors were among the most notable sufferers. (Anyone tempted to picture the disease as purely imaginary is referred to Mandeville's remarkable *A Treatise of the Hypochondriack and Hysteric Diseases*.[52]) In order to compensate for the new wave of sedentary mass culture – if not for higher spiritual purposes – what the net generation needs at school might be something as basic as regular meditation, t'ai chi or yoga.

Progress?

The dawning age of the internet is being greeted with considerable expectation and there can be no doubt that it will bring radical changes. Whether the world on the whole will be a better place as a result – whether technological progress will, in this instance, amount to something worthy of the name of *human* progress – is an open question. The wheel of history is certainly turning. The question we need to ask ourselves is – who finds themselves truly at the forefront of the great revolution? Is it those who make ready use of the great plunging movement of the leading edge of the wheel? Or those who assist in the countermovement that is necessary to raise the back edge of the wheel, however arduously, to the top of its arc?

Cultural resources for coping with technology

The trend towards globalization that has become an almost universal talking point in today's world is certainly not just an economic process. It involves the global diffusion of a technological civilization and hence catches up just about every aspect of human beings' social existence. The question we want to turn to next is how local and regional cultures are faring as today's new world-civilization is taking shape.

The encounter between individual regional cultures and the technical drift of the whole could be developing according to three alternative patterns, which I will summarize in the form of three theses:

1   Technological civilization is undergoing and will continue to undergo a *modification* at the hands of regional culture. (The thesis of *alternative forms of modernity*.)

2   Technological civilization is *destroying* and will continue to destroy regional cultures, whose remains will then provide further material for world-civilization.

3   A relationship of *tension* between technological civilization and the culture of the regions establishes itself over time, leading either to the *division* of life into different cultural or subcultural sectors, or to *resistance* to the technification of the conditions of life.

Despite the theoretical plausibility of all three theses, they also raise a question – something detectable in our adjacent use of the terms *culture* and *civilization* to express our problem in the first place. For example, if, on our second thesis, a technological civilization tends to destroy regional cultures, why not call that a step towards the creation of a world *culture*? Why speak at all of a technological civilization and not a technological culture? Is the preference that many intellectuals have for talking about culture instead of civilization a sign of the characteristically modern repudiation of the artificial, 'external', rationalistic bias of civilization – an expression of the sort of resentment that has been described so well by Norbert Elias and which looks to culture as the touchstone of supposedly more humane, 'organic', authentic human qualities?[53]

So it would seem that we are on tricky interpretative ground here. To avoid the worst pitfalls, we are going to have to give new meaning to the terminological divide between culture and civilization. There are two important reasons for doing so. The first involves a factor in the civilizing process that Elias pays scant attention to: this is the civilizing function of technology itself.[54] What we need to do here is to study processes of technification from a *human* point of view, focusing squarely on the changes to forms of human behaviour and relationship that result from the introduction of new technology. Specifically, what we need to address are new modes of *perception* brought about by photography, film and television, new modes of *communication* brought about by telephones and mobile phones, new forms of *work* that flow from the conversion of all manner of experience and knowledge into information, and new forms of *social relationship* that flow from the formation of new technical networks. One of the fundamental features of all such changes is the way that they relieve behaviour of the need for moral evaluation, a second is the way that they undermine forms of *praxis* in favour either of *poeisis* or purely instrumental action (above all, by giving preference to efficiency over quality). In the West, such processes of technification have already led to a tangible tension between civilization and culture as the former comes to refer to the worlds of work and transport, and the latter comes to refer to the worlds of consumption and leisure.

This leads us to the second ground for reconceiving the difference between civilization and culture: beyond the cultural sphere of the West, we hear the demand for a *de-Westernization* of technology and industry. The intentions of those who propose such a reinterpretation are clear, viz. to open the way for non-Western countries to participate in the benefits of modernization without giving up their cultural identity or inflicting on themselves the destructive side effects of technological civilization so noticeable in the Western world. The Ghanean author Kwame Gyekye is one among many voices in the developing world to insist that modernization should not be interpreted as Westernization even though it involves the successful appropriation of some of Western science and technology.[55] The distinction that Gyekye thinks can be drawn between technified forms of life and a deeper layer of human beliefs – or between technological foreground and background world views – justifies the supposition, I think, that we are living through a period of quite intense conflict between a rapidly advancing technological civilization and individual regional cultures.

The questions we want to pose of this situation thus relate in the main to the third interpretative scheme that we initially set out. Our overarching question will be: what potential for coping with the inroads of technification do the world's existing individual cultures still possess? And what potential do they harbour for resisting those inroads?

A preliminary consideration here is whether we have to answer that question differently for Europe and for other regional cultures. In Europe, after all, the development of modern technology gave rise from the outset to critical confrontations, essentially because European culture was the cradle of modern technology and because the technified world of social modernity took shape there relatively gradually. Other cultures have usually had to confront modern technology in a much more sudden and dramatic way. In what follows, we will try to say in more detailed terms what critical resources for coming to grips with technology Europe has inherited from the past. We can say from the outset that our overall thesis is a sceptical one: the line of thought that we will pursue leads to the conclusion that, in our present situation, *technology itself* is sapping the resources that we might otherwise draw on to criticize technological development. Before proceeding with the argument though, let me illustrate the point with an example.

My example relates to the recent history of childbirth[56] and, in particular, to the rise of so-called *programmed delivery* from the late 1960s on, both in Europe and North America. Childbirth, in short, became a *procedure* carried out under technically optimal conditions using a broad array of scientific data and equipment. It became the norm for labour to be artificially induced and for the course of parturition to be pharmacologically managed and technologically monitored at every step. Birth became a heavily anaesthetized process, taking place often without the active participation or even the consciousness of the mother; caesarean sections became common.

Now of course, one can object to such flagrant medical technification on the grounds that the birth of a new human being *ought to remain a natural event under any circumstances*. In doing so, one would be invoking a *topos*, viz. nature, which has constituted one of Europe's most important cultural resources for resisting the inroads of technological civilization for many centuries. The problem here, however, is that in the context of childbirth, as it came under the sway of medical technology, the notion of nature had become little more than a pale abstraction. As a cultural resource, we might say, nature had largely exhausted its critical potential. More precisely, although it remained possible to argue against programmed delivery on the grounds that birth is *ipso facto* a natural occurrence, the fact that 95 per cent of births were already taking place in hospitals meant that nature had already forfeited its reality as part of the practical life-experience of most women. Hospital birth had become little more than the implementation of a procedure under laboratory conditions in accordance with the technical norms of doctors, nurses and midwives (the latter considered strictly as technical personnel).

Resistance to the concept of programmed birth might have remained purely ideological, had it not been for the proposals of Grantly Dick-Read, the upshot of which was a revitalization of nature's role as a cultural resource. Natural birth, on Dick-Read's picture,[57] consists essentially in a series of practical exercises that aim to give nature new latitude. Apart from prenatal parental education, the key elements of the methodology are flexibility exercises, breathing exercises and learning to adopt a position of relaxed bodily passivity. In the latter case, the critical approach aims to teach mothers to reconceptualize their own *activity* as a kind of *accompaniment* to what comes about naturally of its own accord.

Dick-Read's approach met with considerable success and quickly gave rise to a natural birth movement with broad support from women. Natural birth was influential in changing medical attitudes to birth in hospitals. Far from remaining a marginal initiative, e.g. in support of home birth, it had its effect on the medical mainstream too. Nowadays, programmed delivery is no longer the ideal of the medical profession either.

As encouraging though as the case may be, for our present purposes we need analyse it no further.[58] Instead, let us turn to the more general question of cultural resources for engaging critically with technology, which we will consciously seek to pose from a European point of view.

Critically engaging with technological development: Observations from a European point of view

In his *Philosophical Discourse of Modernity*, Jürgen Habermas[59] has comprehensively shown that the project of modernity in its European incarnation in no way amounts to a straightforward realization of well-formed historical intentions. Habermas maintains rather that European modernity has seen the dominance of some of the Enlightenment's key ideas constantly knocked out of shape by waves of Romanticism and counter-Enlightenment. Yet *The Philosophical Discourse of Modernity* addresses this fascinating circle of problems solely at the level of *discourse*. It overlooks the development of technology and the question of cultural practice, as perhaps one might expect from a work that wants to approach history from the philosophical high plains. My question is a different one to Habermas'. It is this: what cultural forces does the project of modernity have to confront insofar as it takes the form of a mastery of nature, viz. of nature in its dual signification as external nature *and* the nature that we ourselves are? My answer, in sum, will be that there are four broad aspects of our culture that have sustained critical confrontation with the project of technological development: nature, creation, subjectivity and history. Stated schematically like that, they might sound like little more than abbreviated value-judgements. To give them explanatory purchase, our first task is to ask how they are (or were) bound up with specific cultural practices.

The *topos* of nature is one of the most important of all the characteristic marks of European culture as a whole. Since the time of the Greek sophists, *physis* and its many cognates have formed one side of a series of conceptual contrasts that have structured practically all moments of the European world view: nature, as we argued in Chapter 5, has been opposed variously to law and social norms, technology, spirit and civilization. All such oppositions do the strange service of imbuing human life with fundamental ambiguities – each positions human existence somewhere in the middle between two conceptual poles, each as it were divides the basic being of man in two. In the classic instance that we have already touched on many times, nature for instance is what exists of its own accord (thus something that includes human beings partly within its ambit), while non-nature, in the form of technology, is what man himself creates. Man, we might say, is the criterion by which the dividing line between the two sides of the divide is drawn; he belongs on both sides of the divide and will always remain on both sides.

Modernity radically reinterprets that situation though: it places itself unambiguously on the non-natural side of the divide and proclaims the goal of human history to lie in the mastery of nature. The project of freedom is interpreted along similar lines, for in modernity its basic content becomes emancipation from the dictates of natural necessity. Little wonder then that freedom itself is liable to raise hackles wherever it takes on the concrete form of technological development. Resistance of one sort or another is to be found in almost all spheres of society and culture: in education, medicine, agriculture, even in art. If, for example, the Enlightenment placed a premium on inner discipline and the enlightened mastery of inner nature – which in practice meant a mastery of physical need and a repression of bodily impulse – the nineteenth and twentieth centuries saw the formation of a countermovement that sought to foster the spontaneous development of the child and to make possible a progressive expansion of physical sensibility according to the motto of an 'educated nature'. The whole nineteenth-century *discovery of childhood*, as Philippe Ariès[60] calls it, has had a long-lasting influence on all mainstream education, while beyond the mainstream the conception of childhood as a separate phase of human development (as opposed to a mere prelude to adulthood) has given rise to a range of separate teaching styles, the best-known of which are associated with the names of Steiner, Montessori and Waldorf.

In agriculture, the main line of technological development was aimed at maximal appropriation of nature, which in turn meant industrialized farming (so-called 'agribusiness'), the thorough-going mechanization of farm work and the widespread use of industrial chemicals. From the outset, the process met with the resistance of the traditional agricultural economy, which in Europe has hindered the full-scale industrialization of farming down to today. Of course, it is true that the methods of industrial farming – principally the use of high-tech machinery as well as artificial fertilizers and pesticides – have forced their way into traditional agriculture. Yet the twentieth century has also seen the emergence of self-aware alternatives. By early in the twenty-first century, biodynamic farming looks to have established itself as a fixed feature of the landscape, with its own brands and product standards and an alternative marketing system.

In the field of health, resistance to high-tech medicine and the pharmaceutical industry has given rise to naturopathy, homeopathy, anthroposophical medicine and a host of related *alternative* medicines – not just to alternative medical theories or counter concepts, but to the establishment of alternative forms of medical practice and the creation of new healing professions. Public acknowledgement of the vocation of the *health practitioner* is a good example of the broad trend. The role of the German *Heilpraktiker*, who is not formally a member of the medical profession, has over time become an established social institution with recognized courses of study and accreditation and, just as importantly, a recognized place in the health insurance system.

As another form of cultural practice that has consciously sought to defy the mainstream project of the modern Enlightenment in the name of nature, let us turn lastly to art. The classical example of the divide between naturalism and Enlightened control here is the contrast between the English and French garden; while the latter aims to bring the human drive to mastery to a high pitch of aesthetic perfection by careful planning and rigorous geometrical formalism, the goal of the former is nothing more and nothing less than making nature's spontaneous self-activity fully visible. The wilful mastery of nature manifest in the French approach gives way in the English garden to carefully crafted scenes that nature itself might have brought forth.[61]

By contrast, the mainstream of modern art is shaped by the central ideas of social modernity and its ideal of emancipation from nature and that means, first and foremost, a renunciation of the classical principle that art ought to be an imitation of nature. Yet here too counter-tendencies have become apparent. In the art of the twentieth century, we find repeated attempts to consciously work with nature or in nature. One example that comes readily to mind is the *land art movement*, whose aim is simply to make nature visible, as it were by bringing it to a sort of aesthetic articulacy. A host of other artists who have exposed their work to natural weathering or have integrated processes of natural decay and dissipation into the artistic process might also be mentioned here. Artists have variously sought to show us nature in its profusion and destructiveness, as well as in its sparseness and, at times, mind-numbing uniformity. Some have taken to the task of reacquainting human beings in and through art with *their own* nature. Corporeality here becomes not just an abstract aesthetic theme, but a kind of aesthetic practice, the artwork a vehicle for an enhanced sensory experience, or, again, a sort of education of physical sensibility.

Second, we have *creation*. Although closely related to nature as a *topos*, our reason for treating it separately is that its place within the overall constellation of European culture is quite different; while we are indebted to the Greeks for their notion of nature as *physis*, creation is a fundamental part of Europe's Judaeo-Christian heritage. The Biblical picture, in rough terms, is of the cosmos in its entirety, and the earthly realm of nature in particular, as a work of divinity. God – or so the thought goes – arranged the world in a certain way in the beginning and at the completion of the creation process gave confirmation that the product of His creative labours was good. The broader implication for our picture of creation is that the basic constitution of the world is given by God – that the order of the world is indeed something sanctioned by God. Again, man's place within creation, so conceived, is undoubtedly shot through with a basic ambivalence. On the one hand, human beings are themselves one type of creature among others and their divine creator urges them to show dutiful regard for the order of creation that He is the author of. On the other hand, human beings stand out from the rest of the created order; the opening chapter of Genesis (I.28) tells us that man is created in God's image and charged with making himself master of the earth. That ambivalence resonates throughout Western culture down the ages. As the project of modernity was getting underway in the seventeenth century, the mandate that the Bible apparently gave to man's efforts to subdue the world around him was used to legitimate a radical programme to appropriate nature by technological means. On the other hand, a quite different attitudinal complex has evolved out of the notion of respect for creation, in Europe giving rise to numerous forms of cultural practice and to forms of resistance to the limitless technification of both external and human nature, often with the support of the Church. The notion of man's *creaturely* nature is adduced in support of moves to recognize animals as man's fellow creatures. In Germany, the idea has even entered law in the form of legislation[62] preventing cruelty to animals.[63] Relatedly, the Catholic Church has involved itself in militant resistance to technological manipulation of human reproductive life throughout the entire period from conception until birth.

Third, we have the *topos* of *subjectivity* – unlike nature and creation, a value-construct that is a product of modernity without deep roots in the ancient world. (At best, we could point to a premiss of the modern interpretation of subjectivity within the broad nexus of Judaeo-Christian religion: it consists in the notion that God's care extends to every individual human being *whom* He addresses his spiritual call to directly, and whom he also *calls to account*, whom, in Isaiah's words, he 'even calls by name'.[64]) Subjectivity in the sense of an individualized world view determined by the individual human subject's unique experience of the world is certainly only a product of the late eighteenth and early nineteenth centuries, specifically of the French Revolution and European Romanticism. Again, we are talking about a constitutive feature of social institutions and everyday forms of life rather than just an abstract idea. Thus, it is only since the French Revolution that the world has had anything like a concept of individual human rights, only since Romanticism that the individual choice of a partner in marriage has become a customary feature of European culture.

In the world of contemporary modernity, subjectivity – both as a form of life and as a social institution – represents an important base for resistance to the limitless manipulation of human beings. The concept constitutes one of the most important moments of the notion of human dignity. The inviolable status of human dignity demands that respect be shown to individual subjectivity in all dealings with all people. It is precisely such an interpretation of human dignity that today underpins the ban on cloning in Europe and other parts of the world, for to clone a human being would be to seriously encroach on the individual's sense of subjective uniqueness; it would effectively entail doing away with the individual's right to understand himself as a *new beginning*. At a global level, it is man's self-conception as the bearer of individual subjectivity that underpins the UN ban on the equation of individual identity with a sequence of genetic code.

Lastly, the topos of *history*. It, too, is historically relative for the notion of *having a history* or being embedded in history is again an interpretative posit that dates back only as far as the age of Romanticism. It was perhaps only a time marked by social revolutions, industrialization and technification that could confront human beings with the destruction of the past dramatically enough to bring history to light, both existentially and intellectually – only, we might say, an age that called upon human beings to bid farewell to the past that could turn the past into an autonomous source of human value.

Again, we are speaking of history as anything but an abstract locus of human value. We have to remember that living historically gradually becomes an actual form of life, as well as providing the impulse for the creation of a variety of social institutions. In the course of the nineteenth century, clubs and associations sprang up throughout Europe for the express purpose of tending all manner of historical traditions, societies for the cultivation of spoken language traditions and traditional music, for the conservation of regional costumes and, lastly and most importantly, for the protection of local cultural identity – what in German is known as the *Heimatbewegung*. The various European movements to protect local culture were, in many senses, the source of the twentieth-century environment movement,[65] though the concern of the German *Heimat* movement, and its many non-German counterparts, was never with the conservation of nature as such – never, that is, with *wilderness* – but rather with land and landscape as it has been shaped by human history. As a cultural resource, historicity thus implies resistance to a certain sort of arbitrary technological tinkering with natural and cultural givens. Historicity demands that innovation show due respect for tradition, and that the new be integrated harmoniously into the old. That means respect for historical monuments and the built environment as well as the natural environment. In Germany, both types of conservation are now anchored in law at a state level. Conservation of nature – not of nature as such, but of nature as an inalienably *human* habitat – has, in brief, become one of the prerogatives of government.

Since the nineteenth century, several major social movements have drawn on historicity as a cultural and environmental resource and, in the form of various associations and interest groups, they remain active and effective down to today. In comparison to the other three *topoi* that we have examined, historicity indeed seems to open out one possibility that makes it particularly useful as a means for coping with technological development. When we say a form of life is shaped by history, that by no means implies that it need simply be seen as a touchstone of *resistance* to innovations. Rather, we mean that what is to be preserved should be seen as a product of human labour and activity. Nature, for instance, can be understood quite non-metaphorically as a cultural landscape and in invoking it as a value we can think of ourselves not as repudiating all change, but as making a positive claim on behalf of cultural and environmental continuity.

And yet when we enumerate all four sets of resources for coping with technological change, it seems questionable whether they could ever find application throughout the world. In exploring each of the four, I have drawn attention to the particular place that each came to occupy within the overall constellation of European cultural history. We might well doubt whether there are functional equivalents of our four *topoi* within other cultures, which could provide analogous starting points for reflection and action. A concept of nature, for instance, first came into existence within the Greek *polis* in contradistinction to a highly developed sense of free human self-determination. As a *topos*, nature was to experience a kind of renaissance in the age of Rousseau, when modern civilization was first felt to be a demoralizing source of complexity and constraint. The interpretation of the world as a specific kind of creation is likewise the expression of a profoundly Judaeo-Christian mentality, and coeval with the Greek interpretation of nature, if not actually more ancient. *Subjectivity* came into its own in the great *melée* of bourgeois political revolution that engulfed Europe in the eighteenth and nineteenth centuries, while *historicity* took shape around 1800 in reaction to the Industrial Revolution. Taken together, all four of our *topoi* – nature, creation, subjectivity and historicity – make up a sort of cultural frame within which Europe's technological development took place. Each of the four had an effect in forcing technological development into particular channels, each drew out the process, each presented obstacles – in short, each played a part in shaping the course of technological history. The result in the long term has been that contemporary technoscience has to abide by laws for environmental and animal protection; that the technological innovations of industry are subject to the scrutiny of government agencies, which evaluate them on their environmental and social merits; that technological innovation in medicine has to heed the demands of human dignity; that scientific experiments are subject to rigorous approval procedures. The regimen of checks and balances that we have put in place functions well in some circumstances, less well in others. However, the question raised by the global spread of technological civilization is not only whether the four sets of European resources for dealing with technological progress are applicable throughout the world, nor whether other cultures possess functional equivalents of our four major *topoi*. Instead, it is the question of how far and how long we can rely on these resources. For if the resources themselves are being depleted, our interpretative as well as our practical situation changes radically, both in the West and beyond.

Critical attrition

My sceptical thesis is that the resources that Europe has historically had at its disposal for coping with technological progress are themselves being slowly destroyed. Why that is the case goes to the heart of why I am inclined to call contemporary civilization a *technological* civilization. Over the past century, technology has come to occupy a pre-eminent position within the overall edifice of civilization; traditional cultures, in Europe and beyond, have been marginalized or extinguished in the process. Our world no longer presents a picture of individual civilizations or cultures that make use of particular technologies. We must reverse the order of explanation as it were: particular technologies have themselves taken on a formative function throughout the whole of human existence and social life. The transformation that we have in mind could be described under the rubric of a technification of all areas of life. Moreover, if we want to use the concept of technification to grasp contemporary developments, it immediately becomes clear that the process reaches well back into the past and perhaps simply makes explicit and irreversible what was implicit in modern technology from the beginning.

Armed with this rather grim suspicion, let us turn again to our four *topoi* and see what more can be said in illustration of the thesis that technological development is eating away at the critical potential of existing cultural resources.

The process is most clearly evident in the field of nature. It hardly needs to be pointed out that nature is something that we readily invoke, especially in Europe, in the search for alternatives to the technification of life, or when we are looking to check the course of technological development. But are we clear what we mean by nature here? Is nature not precisely what we have been studying under experimental, i.e. technological, conditions ever since the time of Galileo? Has knowledge of nature not been equated since Descartes with the technological reproduction of nature's works? Did we not cease long ago to think of nature as what was naturally given and begin to think of it as including what we ourselves produce by technological means, hence as a concept broad enough to include transuranic elements and complex artificial polymers? The further we succeed in reproducing nature technologically – for instance in the corporeal domain that has become the domain of medicine and biotechnology – the less obvious it becomes why we should draw any sort of boundary between the natural and the artificial, and less obvious still why we should respect any such boundary. As the boundary between the natural and the artificial shifts and blurs, the more difficult it becomes to know what resistance to the scientific manipulation of species or to artificial organ transplants or to arbitrary human intervention in established ecosystems should base itself on. If nature is no longer simply, as the Greek notion of *physis* suggests, what shows itself of its own accord and thus what man must take as given, if it instead becomes what can be fabricated in accordance with the underlying laws of nature, then it is clear that the *topos* of nature forfeits its normative meaning. In short, nature ceases, under such circumstances, to be a region of existence that is to be respected and carefully tended. Culturally, it loses its function as a regulative ideal that sets limits to technological developments.

Something similar applies to the *topos* of creation, which has lost its savour as the role of the Church in shaping world views has gone into decline. As the Church has been forced to cede to science all claim to define what nature is, the concept of creation has been subject to a gradual but nonetheless drastic devaluation. The essential point of Christian teaching on this score is that the process of creation, in classical Biblical terms, is *completed on the seventh day*; however, the pressure that science has exerted on Christianity over the course of centuries has compelled the Church to revise this concept of a created order of being from the ground up. Genesis has been reinterpreted to make it compatible with the theory of evolution and with striking results. Apart from the blossoming of 1,000 tropical antiscientific flowers within American Christian fundamentalism, theological possibilities once regarded as heretical have made a comeback, prime among them the pantheist equation of God with nature and the notion of a *permanent* or *ongoing* creation. Strongly influenced by the philosopher Alfred North Whitehead, such ideas have gained increasing acceptance under the heading of process theology, at least in the moderate Protestant sector of the Christian world. However, the problems that the new theology poses should not be understated. With the falling away of the notion of a fixed order of creation that the Christian is bound to respect, the concept of nature loses all firm outlines. If nature is a byword for an ongoing process of creation, then nature will surely continue to develop down the ages. Just as surely it will continue to *be developed* by human beings.

If the religious resources for opposing one-sided technological development have been eroded because of a certain sort of scientification, viz. as a consequence of the way that theology has had to *adjust* to science, then the erosion of subjectivity as a source of potential resistance results from the way that technology itself intrudes into the cultural domain. Invasive technification takes many forms here. Psychotropic drugs have made the realm of human emotion – i.e. precisely what is constitutive of subjectivity qua sensitivity – into a manipulable = X. The identity of individual human beings, which was once grounded culturally in their responsibility for the unity and continuity of their lives, could soon be reduced to a bare biological fact as methods of genetic fingerprinting make rapid headway. On a different front, the possibilities for creatively multiplying personal identity continue to expand as the technological means of producing a multiplicity of data-based and image-based realities swell. Life online becomes a part of real life for more and more people; the loosely connected, quasi-fictional worlds of the internet generate a wild profusion of partial personalities whose coherence we see ourselves increasingly as neither obliged nor able to maintain. A range of cultural practices flourish, which make it increasingly pointless to mobilize resistance to a technified life by invoking an ideal of subjectivity whose integrity is to be respected. We have already mentioned many examples: the proliferation of virtual worlds in online games, the use of psychotropic drugs for therapeutic and non-therapeutic purposes, the use of genetic fingerprinting and the creation of data-storage systems to house the associated reams of raw biological information.

Finally, a word about the *topos* of history. Here, we have a famous fictional example of how history can be obliterated. Orwell's novel *1984* presents us with a world in which history is constantly being rewritten and the body of existing facts about the past moulded to suit daily political requirements. Although the picture is intentionally overdrawn in line with Orwell's political point, it still brings out a host of worrying technological possibilities that are far closer at hand in today's world than they were either at the time that Orwell was writing or in the non-fictional world of the mid-1980s. In specific terms, there is the obvious problem that photographs have lost their value as historical documents as the possibilities of technological manipulation have multiplied. Yet technological loss here does not just bear the negative sign of a past that can be arbitrarily rewritten or extinguished. It manifests itself, perhaps even more insidiously, in the image of a past that is held up to the present in its undifferentiated entirety. As a functional form of cultural practice, history calls for both forgetting and remembering, as well as an element of *tradition*, whereas electronic databases make it possible to give a permanent presence to any and all aspects of the past. Thus, either negatively or positively, the circulation and use of technified images and texts renders the past weightless in a certain sense, robbing it of its gravity and much of its potential as a locus of resistance.

Conclusion

We have tried to show that European culture, though in a sense the wellspring of technological modernity, also has the critical potential to come to terms with technological change. The picture that technological modernity presents us with 400 years since its inception is astonishingly ambivalent: it has led to both a tremendous intensification of human existence and, at the same time, to the very real annihilation of humanity – nothing less than a qualitative destruction of what it once was to be a human being. I have attempted to give an overview of the resources that stood at the disposal of European culture and of the cultural practices that might help redefine and, in a sense, shore up our historical ideal of humanity. I emphasize that I am far from keen to recommend the critical possibilities that have evolved within the framework of European culture to the wider non-European world, particularly given how sceptical I am about whether Europe's critical resources will be enough to withstand all the multiple, and widely divergent, strains of contemporary technification.

It would be remiss to conclude without some remarks about the alternative critical resources that are available within non-European cultures, though the topic would be well worth a fuller treatment. The few comments that I am able to make here on the basis of preliminary intercultural discussions will at least bring into relief the home-grown European resources that I have discussed in detail above. For our present purposes, we can limit ourselves to the *topos* of nature.

The first point to note is that the idea of nature plays a far smaller role in Asian cultures than it does in European culture. Translating the very word into Chinese or Japanese is already a tall order.[66] Thus, for instance, while there are certainly expressions corresponding to the word *naturally* (in the sense of *obviously* in its many cognate European forms), a term for nature as an overall complex (physical nature, the natural world, etc.) is something that Asian languages generally lack. The linguistic gulf is indicative of a wider cultural difference. And the difference becomes all the more striking if one asks how, say, the Japanese or Chinese go about referring to the sorts of concepts and events, places and spaces that Europeans denote using the notion of nature or the natural. In coming to grips with the medical manipulation of the human body for instance, Europeans customarily remind themselves that their bodily existence is essentially *natural*; the problem posed by technological intervention is formulated as a question about a problematic step towards *artificiality*. The Asian experience of Western techniques of organ transplant puts this way of looking at things in perspective.

One of the fundamental presuppositions of organ transplantation is a Western definition of death in medical terms as *brain death*; in recent decades, it has been adopted as part of Japanese law for instance. Nonetheless, transplants are an extremely rare occurrence in Japan, in large part because the majority of Japanese people refuse to accept the definition of death as brain death. Death, for the Japanese, is a continuous process whereby *chi* – here meaning something like 'life force'[67] – returns to the earth. Because the dissipation of *chi* is an *ongoing* process, it makes no sense to speak of a definitive moment of death. A further reason for rejecting Western techniques of organ transplant is deduced from Confucian thought, which requires that the dead body be delivered intact to the realm of the ancestral spirits.[68]

One begins to see how different a hue these sorts of arguments take on in the absence of our customary European discourse of the natural. As far as intercultural communication goes, things become even more awkward when, for instance, one's Indian colleagues raise objections to germ line therapy (which involves medically interfering with a human being's genetic make-up) on the grounds that it contravenes the Hindu doctrine of *kharma* (medical intervention, on this argument, would be disallowed because it would involve altering an individual's basic kharmic disposition). Of course, one needs to ask here how widespread this specific interpretation of *kharma* is among the broader Indian population and to what degree it shapes thought and action in the real world.

We have insisted at various points of the current chapter that debates about technology can never revolve merely around terminology or concepts and that the truly important questions here are questions of cultural practice for two reasons: first because it is cultural practice that provides the essential background to all forms of opposition to the technification of human life, and second, because cultural practice is an indispensible point of reference in negotiating rules relating to the application of technology. In short, cultural relativities must be kept firmly in view here. Universal notions of human rights are unlikely to suffice as a basis for practical, cultural or ethical discourse as individual societies seek to create laws to govern the use of technology, on the one hand because such notions of generalized right are too abstract to provide a basis for people to cope with technological change, and on the other hand because they have already come in for criticism from the point of view of non-Western cultural traditions.

APPENDIX

The last man as Übermensch

Nietzsche

And thus spoke Zarathustra to the people:

It is time for man to fix his goal. It is time for man to plant the seed of his highest hope.

His soil is still rich enough for it. But this soil will one day be poor and weak; no longer will a high tree be able to grow from it.

Alas! The time is coming when man will no more shoot the arrow of his longing out over mankind, and the string of his bow will have forgotten how to twang!

I tell you: one must have chaos in one, to give birth to a dancing star. I tell you: you still have chaos in you.

Alas! The time is coming when man will give birth to no more stars. Alas! The time of the most contemptible man is coming, the man who can no longer despise himself.

Behold! I shall show you the *Ultimate Man*.

"What is love? What is creation? What is longing? What is a star?" thus asks Ultimate Man and blinks.

The earth has become small, and upon it hops the Ultimate Man, who makes everything small. His race is an inexterminable as the flea; the Ultimate Man lives longest.

"We have discovered happiness," say the Ultimate Men and blink.

They have left the places where living was hard: for one needs warmth. One still loves one's neighbour and rubs oneself against him: for one needs warmth.

Sickness and mistrust count as sins with them: one should go about warily. He is a fool who still stumbles over stones or over men!

A little poison now and then: that produces pleasant dreams. And a lot of poison at last, for a pleasant death.

They still work, for work is entertainment. But they take care the entertainment does not exhaust them.

Nobody grows rich or poor any more: both are too much of a burden. Who still wants to rule? Who obey? Both are too much of a burden.

No herdsman and one herd. Everyone wants the same thing, everyone is the same: whoever thinks otherwise goes voluntarily into the madhouse.

"Formerly all the world was mad," say the most acute of them and blink.

They are clever and know everything that has ever happened: so there is no end to their mockery. They still quarrel, but they soon make up – otherwise indigestion would result.

They have their little pleasure for the day and their little pleasure for the night: but they respect health.

"We have discovered happiness," say the Ultimate Men and blink.

And here ended Zarathustra's first discourse, which is also called "The Prologue"; for at this point the shouting and mirth of the crowd interrupted him. "Give us this Ultimate Man, O Zarathustra" – so they cried – "make us into this Ultimate Man! You can have the Superman!" And all the people laughed and shouted.[1]

In Zarathustra's grand disquisition on 'the last man', Nietzsche sketches a kind of horror scenario: he extrapolates the direction he can see humanity heading in and has already censured as *decadent* in the severest terms. The type of human being that he is imagining is one whose foremost concerns are security and pleasure, a type whose understanding of morality is purely utilitarian; the last man, we might say, using our quasi-analytical terms rather than Nietzsche's quasi-prophetic ones, takes *technical and economic* progress as the sum and substance of *human* progress. As is well known, Nietzsche's counter-image to the last man is the *Übermensch*, the figure who in *Thus Spoke Zarathustra* becomes the centrepoint of a vision of the human future in which man consciously advances beyond his human all-too-human limitations in an act of sternly jubiliant self-overcoming:

Man is a rope, fastened between animal and Superman – a rope over an abyss.[2]

The remainder of *Zarathustra* makes it clear that Nietzsche thinks of the project of self-overcoming as an exercise in self-cultivation – as a moral and aesthetic intensification of existence. The character types that he clearly takes his bearings from in conceiving the *Übermensch* are the artist, the free spirit, the heroic individual who risks his life.

What are we to make of it all in the present? More than a hundred years after Nietzsche, we find ourselves caught up in a very different project of human self-overcoming. Our project, too, involves a comprehensive step beyond the human all-too-human, an attempt to bravely cast aside a host of human dependencies, particularly dependence on nature. Yet the cultural ideals that provide the guiding stars of our contemporary line of development are quite different – on the one hand, media-generated visual idols, and on the other hand, figures like the cyborg and the transgenic human being. The means increasingly at our disposal to reach these heady goals are different too; our means are *technical* means placed in the service of a search for technical perfection that has nothing whatsoever to do with the aesthetic or moral strenuousness favoured by the author of *Thus Spoke Zarathustra*.

From Nietzsche's point of view, we might almost say that the ideal image projected by the last man of today *is* that of the *Übermensch*. Both the contemporary cult of the body and contemporary biotechnology illustrate this well enough. In the present essay, I have chosen to leave both those topics aside and focus instead on a different kind of technology – transplantation technology. My rationale here is that we have a philosophical text that explicitly relates the technology to Nietzsche's vision of the *Übermensch*. That text is the philosopher Jean-Luc Nancy's *L'Intrus* (*The Intruder*).[3]

When he wrote *The Intruder*, Nancy had already lived for 10 years with a transplanted heart. His essay brings together his at times exceptionally painful experiences, situating them in the context of a broad series of reflections about human life that have developed continuously from the philosophical anthropology of the ancient world, via Nietzsche, down to the present day.

Jean-Luc Nancy

Nancy was diagnosed with cardiomyopathy in the 1980s, and underwent major surgery in 1990. Progressive cardiac insufficiency of no known cause effectively meant that without a transplant he would have been unlikely to live beyond his early fifties. Obviously, the choice between the transplant and refusing treatment hardly existed at all; as is so often the case in modern medicine, the patient found himself involved in a course of action that he experienced as a kind of *fait accompli*. A little way into the text, we read: 'Each of these people [the medical staff] came to the conclusion for his own reasons that it would be worth the trouble to prolong my life. It is not difficult to imagine the complexities, the interplay of delicate considerations, which are brought to bear in these cases by *others* and which reach into the depths of one's own living being'.[4]

What follows is Nancy's account of his ordeals, the crux of which is that he experiences the transplanted organ as a kind of intruder – a foreign body that makes him feel ill at ease with himself. *L'Intrus* interweaves reflections about his condition with details of the phase after the transplant, a story of bitter suffering. Post-operative life is necessarily one of strict medical supervision and extremely limited activity – Nancy's life is constantly under threat from fresh outbreaks of disease, in particular from viral infections. In the eighth year after the transplant, he develops cancer. A longer excerpt will be enough to give an impression of the medical martyrdom that the philosopher's life descends into:

> From one round of pain to another, from one unnerving experience to the next, in the end *I* am nothing but the thinnest of threads. One reaches a stage at which the repeated medical intrusions form a comprehensive whole, a never-ending regime of intrusions. Added to the routine daily doses of various medicines and the barrage of tests, there are the effects of the radiotherapy on teeth, the incurable dry mouth, the quarantining of every mouthful I eat, of every potentially infectious contact with the world, the wasting of muscles and kidney failure, the loss of memory and the loss of interest in work, the constant poring over test results, the insidious recurrence of respiratory tract infections, the candidosis, the polyneuritis and the general feeling of being indistinguishable from a vast web of procedures and observations, from chemical and institutional signals which one neglects at one's own peril as if they were the very threads that everyday life were woven out of, but which in reality simply seem to remind one of their own existence and of the need for rigorous control. I am at the point where I am inseparable from a multiform solution to a set of medical problems. Maybe the lives of the sick and the elderly have always looked like this more or less. And yet I am neither simply a sick man, nor merely an old man. What I am being healed by [. . .], what keeps me alive, rapidly accelerates aging. My heart is 20 years younger than me, the rest of my body is a dozen years older at least. Simultaneously younger and older than my years, no precise age is right for me, or even relevant to me – actually I am ageless. I don't have a real job anymore, though I haven't retired. Nothing remains of what I should be, I am neither husband, nor father, nor grandfather, nor friend, or rather I am these things conditionally, viz. on the general condition of the intruder, in fact on the condition of a host of intruders which can take my place at any moment in my relations with The Other or in The Other's mental picture of me.[5]

How, on balance, does he judge what he goes through? Was it worth the trouble? It will astonish many readers that Nancy's tone at the end of his depiction of such senseless suffering is neither one of resignation nor despair. Instead of any clear-cut repudiation of any aspect of his experience, we find the following almost triumphal assessment:

> Along with all those fellow-sufferers who grow ever more numerous, I might actually call myself the first sign of a certain mutation. Humanity is starting out on the path (as it does time and again) of endless self-overcoming (for, however

one interprets it, this is the most basic meaning of the "death of God".) Man is in the process of becoming what he is – the most terrifying and unsettling of all technicians (precisely what Sophocles described him as 25 centuries ago), the being who robs nature of her naturalness in order to create her anew, who re-invents the whole of creation, who induces it to appear from out of nothingness and who will perhaps consign it to nothingness again. We might say that man is on the way to becoming the being who holds the beginning and the end in his creative hand.[6]

The reference points in Nietzsche are two of the best known: the first is Zarathustra's *Übermensch* (the figure that the prologue to *Zarathustra* tells us man should be considered a bridge to), the second is section 125 of *The Gay Science*, 'The Madman', whose eponymous central figure proclaims the 'murder' of God to be so great a deed that in its aftermath human beings are bound to lose all sense of existential orientation. Nancy's other reference is to the first choral stasimon of *Antigone*, Sophocles' great song of praise to man the great artificer and master of nature, 'There are many strange and wonderful things, but nothing more strangely wonderful than man. . .'. But it is surely the tone of (philosophically amplified) exuberance that makes for most interesting reading – a strange tone to strike for a man in great distress who has just described the wretchedness of his existence with remarkable clarity.

How are we to understand Nancy's reaction? Perhaps as a heroic self-stylization on the part of a man in despair? As the expression of a belief in technological progress that is so unshakeable that it writes off catastrophic failure as a temporary imperfection of the technical means at our disposal? As age-old utopian indifference to the number of human victims of technological dreams? Or as a kind of mask of a personal, or rather a philosophical, failure?

The problem of identity

Nancy's choice to analyse the existential problems posed by his heart transplant as problems of the self and the other is no doubt the right one. In fact, it is difficult to imagine a better way of sheeting home the philosophical issues surrounding contemporary medicine in equally personal *and* conceptual terms. What is more dubious though is the analogy that he draws between his experiences and the political issue of migration. In clear reference to Simmel's reading of the figure of the foreigner,[7] Nancy writes:

Once he has arrived, if he remains foreign, and for as long as he does so – rather than simply "becoming naturalised" – his coming will not cease: nor will it cease being in some respect an intrusion.[8]

Nancy appears to be making an implicit call here for rigorous submission on the part of the foreigner (the other) to the regime of the new homeland (the self). And perhaps the instrumental rationality that underpins modern medicine gives his case a certain plausibility; certainly in the conventional technical understanding of organ transplantation, what major operations require of the human body is a policy of rigorous assimilation. Yet in the mirror that the political analogy holds up to medical rationality, a critical light falls across Nancy's own coping strategy too. One wonders whether coming to terms with the transplant might have been possible in other terms – whether the political concept/metaphor might have been troped rather differently. What of the possibility of simply *letting the foreigner/ other be*? In a footnote to the text, Nancy makes mention of a drawing by Sylvie Blocher (a friend) with the title 'Jean-Luc and his woman's heart'. It would certainly have helped to see Blocher's piece included in the text of *The Intruder*. One suspects from the title alone that it suggests quite different interpretations of the self/other relation from Nancy's own.

The existential problem of organ transplantation can be cogently framed using the interpretative category of the foreign only as long as one presupposes that identifying with one's own body is quite basic to one's individual mode of being, and it is this presupposition that Nancy fails to fulfil. Even before his operation, his own heart seems to have been something of a foreign object:

> If my heart was giving up on me, to what degree was it an organ of "mine", my "own"? Was it even an organ?[9]

It seems right to read *The Intruder* at this point through the Heideggerian notion of conspicuousness [*Auffälligkeit*]; here and elsewhere in the text, one detects the influence of Heidegger's theory of equipment in Nancy's way of formulating the problem that his experience poses for him.[10] Heidegger, as we have noted before, holds that an instrument, in this case the heart considered purely as a bodily organ, is *inconspicuous* precisely when it performs its normal function. It is only when it fails to do so that it announces itself as present, becomes *con*spicuous.

As interesting as this may be philosophically, what Nancy's Heideggerian reading of his post-operative predicament resoundingly fails to suggest is any sort of ability to identify with his heart before the onset of his illness. Philosophically a rigorous Cartesian, Nancy emphasizes precisely his capacity for disidentification with his body:

> Not "my heart" endlessly beating, as absent to me till now as the soles of my feet walking . . . It was becoming a stranger to me, intruding through its defection – almost through rejection, if not dejection. I had this heart somewhere near my lips

or on my tongue, like an improper food . . . a sort of mild indigestion. A gradual slippage was separating me from myself. There I was: it was summer, I had to wait, something was detaching itself from me, or was coming up in me, there where nothing had been: nothing but the "proper" immersion in me of "myself" that had never identified itself as this body, even less as this heart . . .[11]

The French original makes the point even more graphically than our English translation: immersion in the ebb and flow of his own bodily nature has never been a part of Nancy's sense of self, so that, initially at least, transplanting another's heart into his own body poses no existential problem for him.

What becomes clear in the aftermath of surgery is that his initial reaction was the result of a kind of existential miscalculation. Nancy begins by assuming that his sense of self will be completely unaffected by his operation because his corporeality seems quite inessential to who he is. What he ends up having to cope with is the self-estrangement produced by the intrusive heart of another. As he himself put it, 'the intruder takes up position inside me: I become entirely foreign to myself'.[12]

This is the problem in a nutshell: following the operation, Nancy is confronted for the first time, *ex negativo* as it were, with the fact that his identity as a self has a bodily substratum; the post-operative bodily tumult forces on him what Horkheimer and Adorno refer to as a 'remembrance of the nature that subsists in the subject'.[13]

The primary manifestation of the foreignness of his new heart is his severe immune reactions to the transplant:

What happens in any case is that identity and immunity come to coincide. A weakening of one's immunity means a weakening of one's entire identity.[14]

Together with the belated recognition that his own identity has a corporeal basis, Nancy is forced to recognize that the latter is undermined, one might almost say annulled, by his body's post-operative release of immunosuppressants. Psychologically, one could say that Nancy suffers a serious diffusion of identity[15] as a kind of secondary illness. The consequences are drastic, for the missing self-identification with his own body is not something he can make good retrospectively; identity *qua* immunity in fact turns out to be yet another thing that he experiences as foreign. The text speaks paradoxically of the 'foreignness of my own identity'.[16]

The ego that he conceived of initially in terms of a moment of pure self-identification disintegrates into an *I* and a *me*:

the empty identity of an "I" can no longer be based either on any straightforward correspondence with or adjustment to my body or on formulaic self-identification of the kind $I = I$. The verbal form "I suffer" contains two I's which are foreign to one another, though they nonetheless come into contact with one another in some ways. The same applies to the verbal form "I experience pleasure", "I feel good", as it would be possible to show from the use of both forms of expression. In the formula "I suffer", one "I" disowns the other, while in the phrase "I experience pleasure" the self transcends the other. The two verbal forms are as alike as two drops of water – but that is as far as the resemblance goes.[17]

Obviously, we could say that Nancy's prolonged sufferings give him a chance to experience his corporeality retroactively through a process of affective immersion in the vulnerabilities of the self. As it happens, his illness arrives too late. Nancy's practical sense of himself remains quintessentially Cartesian. The emergence of a bodily sense of self turns out to be something he can only experience as self-alienation.

The moral problem

The sole question that allows Nancy to approach his experience of the transplant from a moral angle turns out to be the question: was it all worth it? And we ought to bear in mind that the French *est-ce qu'il vaut la peine?* gives the question a special sharpness: the French word *peine* is literally a synonym of pain rather than a generic term for trouble or difficulty. The really important point is this though: in the form that Nancy raises the question, it remains a question of utility, not a moral question in the full sense of the word. In short, it leads Nancy to weigh the pros and cons of prolonging life. What it does not bring into focus is the kind of human being that he becomes by choosing one course of action over another. If it is right to say, as we have argued at length elsewhere,[18] that moral situations are defined by a certain seriousness – the seriousness of a choice between becoming one type of human being or another – then it would seem that Nancy fails to take the situation that he finds himself in with sufficient seriousness. His reason for submitting to drastic medical intervention amounts to nothing more than a desire to prolong life for its own sake, and yet in light of the question *est-ce qu'il vaut la peine?* that reason strikes him as insufficient. As the post-operative phase of the experience unfolds, the more his life comes to consist in pain and suffering, which are the price he has to pay for his continued survival. And so the question he increasingly asks himself is: 'Why is long life considered to be a good?'[19]

Had he been inclined to doubt whether it was such a good, he might have turned to Plato's famous picture in the *Gorgias* of the helmsman who has brought his passengers safely ashore after a perilous sea journey, and who now strolls up and down a beach contemplating the fact that he doesn't know which of them it would have been better not to rescue.[20]

Nancy doesn't take up the thread until it is too late, and as a result discovers, to his consternation, that prolonging life for its own sake actually empties it of meaning:

> At least since Descartes, modern human beings have pressed their desire for longer life and immortality into the general service of mastering nature. They have thus paved the way for growing alienation from nature, and in the process roused the absolute strangeness of the paradox of mortality and immortality to new life. The problem of life and death which religion took it upon itself to answer is something they try to solve by technical means. The latter however simply postpone the end of life and thus defer the question of life's purpose. By prolonging life in a mere temporal sense, technology highlights its endlessness and aimlessness, its lack of purpose. So we should try to live longer lives – to what purpose? So we can put off dying – but that also means putting death on display, drawing attention to it [. . . .] One must realise that humanity has at no point been equipped to answer this question in any of its various forms.[21]

Now it becomes clear in the course of his reflections that Nancy too is ill equipped to answer the question, and that he reproaches himself as a philosopher for being unable to do so. On his own account, the heart transplant is something that he passively accepts; every step he takes down the path of further medical intervention follows a predetermined pattern dictated by medical technology. His passivity is understandable – many human beings will have had similar experiences. Yet one still wonders whether more might have been expected from a philosopher. Who, if not a philosopher, is going to live up to the ideal of the informed patient exercising a mature capacity for choice? Or, in more traditional terms, is preparation for death not supposed to be one of the prime moments of a philosophical way of life?

In conclusion, let us try to sketch an answer to the question what such preparation could amount to in the context of contemporary medical developments. On what basis can the question whether to submit oneself to organ transplantation, particularly a heart transplant, be decided? Looking back at Jean-Luc Nancy's painful story and at his inability to come to terms with that story morally, there is one basic ethical conclusion that we can draw, a point that looks almost revolutionary against the background of previous understandings of ethics.

There is a widely held view that ethics is essentially a matter of argument, of justification, of principles and decisions. Yet Aristotle had already taken a stand against that view in his *Nicomachean Ethics*, where he criticizes the Socratic identification of virtue with knowledge and argues that ethics can never be solely a matter of ethical discussion or intellectual activity. Ethical life, according to Aristotle, demands practice as well as the ongoing formation of character. If we are to be even remotely equal to moral challenges in critical situations, a sort of ethical *exercise* is necessary. The sort of practice that Aristotle has in mind involves much that is still thought to be vital to moral education today, things like the basic practice of self-mastery, learning to bear a certain level of physical pain, resisting temptation (non-moralistically construed). Yet in our present-day ethical position, given that one of the greatest moral challenges in our time is presented by the possibilities of technologically manipulating the human body, a quite different set of exercises has to be added to the classical Aristotelian curriculum – exercises that make possible a sort of *trust in the givens of the self*, exercises, above all, which enable us to identify with our own bodies and help us learn to truly *live with* our bodies. In a way, it is far from accidental that the richest insights of the philosophy of the body – the notions of the corporeal self, the notion of *betroffene Selbstgegebenheit* ('affected self-givenness' as it has awkwardly been called) – have until now had next to no resonance beyond tiny circles of specialist researchers. In practical terms, however comprehensively it may have been unpicked, the Cartesian view of the body still holds sway. People, on the whole, treat their limbs and internal organs as instruments, their bodies as independent of their essential selves.

Here we come to the nub of the problem with organ transplantation. On the basis of the prevailing view of the body, how am I to decide whether I am prepared to exchange *my heart* for another? Being able to decide surely presupposes that I truly feel my heart to be mine in the first place. Yet if I have lived life identifying with my heart in this way, then I can hardly view it as a mere organ, say as a kind of pump. My heart becomes rather a kind of bodily island,[22] something through which I sense my own vitality and vulnerability, my joy and pleasure, my despondency and, especially, my love. What is normally considered symbolism or metaphor reveals itself as a lived reality. When the reality of one's physical heart takes on all the colours of lived experience, replacing one's heart with another may not be an easy or obvious step. *Purely* for the sake of prolonging one's life, it might well be a step few would take.

Of course, it is impossible to anticipate the decisions that individuals might make, let alone prescribe the decisions that they ought to make. Each human being, faced with a situation similar to Nancy's, will have to make a decision with due consideration of his overall life circumstances, though it can certainly be said, if our view of the quintessential gravity of moral choice is right, that the outcome of one's decision will make a basic difference to the sort of human being one is.

The outcome of such decision-making will also vary depending on the organ that is to be transplanted. This much, however, ought to be clear: if thought is not given to one's lived sense of one's body, the question of whether or how to proceed with a major organ transplant cannot even be posed in adequate moral terms. An ability to stand firm morally, we might say, on this view presupposes having *found a place to stand within oneself*, viz. as a bodily being.

One comment by way of conclusion. It certainly cannot be taken as obviously justifiable to criticize a human being whose years of horrendous pain ought to merit compassion and respect. Such criticism seemed necessary to us because Jean-Luc Nancy's life has been laid open to view as a kind of *experimentum crucis*. To honour the man who has lived out the cruel paradoxes of our contemporary view of the body has been part of our purpose. For the fact that he has presented us with such a remarkable testament of his sufferings – for giving us *The Intruder* in spite of those sufferings – we owe him a debt of gratitude.

NOTES

Chapter 1

1   Helmut Dubiel, *Tief im Hirn*. Munich: Antje Kunstmann, 2006. p. 86.
2   A characteristic example is provided by the philosopher Jean-Luc Nancy, whose implanted heart he describes as an intruder. See Jean-Luc Nancy, *L'Intrus*. Paris: Galilée, 2000.
3   op. cit. Dubiel, p. 92ff.
4   Another fine and by now classic example is the development of a feudal society in response to the invention of the stirrup. See Lynn White Jr., *Die mittelalterliche Technik und der Wandel der Gesellschaft*. Munich: Heinz Moos, 1968.
5   Jacques Ellul, *The Technological Society*. New York: Vintage Books, 1954.
6   Norbert Elias (Trans. Jephcott), *The Civilising Process*. Volume 2, *Power and Civility*. New York: Pantheon Books, 1982.
7   Ibid. p. 223.
8   Gernot Böhme, *Ethik leiblicher Existenz*. Frankfurt am Main: Suhrkamp, 2008.
9   Arnold Gehlen, *Der Mensch. Seine Natur und seine Stellung in der Welt*. Frankfurt am Main: Athenäum, 1962.
10 Ernst Kapp, *Grundlinien einer Philosophie der Technik. Zur Entstehungsgeschichte der Cultur aus neuen Gesichtspunkten*. Braunschweig: Vieweg, 1877.
11 Such was the main thesis of the Darmstadt University 'Technification and Society' colloquium.
12 Bruno Latour, *We Have Never Been Modern*. Cambridge, MA: Harvard University Press, 1993.
13 Donna J. Harraway, 'A cyborg manifesto. Technology and socialist feminism in the late twentieth century', in *Simians, Cyborgs and Women: The Reinvention of Nature*. New York: Routledge, 1991. pp. 149–81.
14 Gernot Böhme, *Am Ende des Baconschen Zeitalters. Studien zur Wissenschaftentwicklung*. Frankfurt am Main: Suhrkamp, 1993.
15 The reviews collected in Christoph Hubig, Alois Huning and Günther Ropohl, *Nachdenken über Technik. Die Klassiker der Technikphilosophie* (Berlin: Edition Sigma, 2000) give a good overview.

16 Aristotle, *Physics* B1. See, for example, http://www.kennydominican.joyeurs.com/GreekClassics/AristotlePhysics.htm.

17 Martin Heidegger (Trans. Macquarrie and Robinson), *Being and Time*. Oxford: Basil Blackwell, 1987. p. 95ff.

18 op. cit. Gehlen.

19 Serge Moscovici, *Essai sur l'histoire de la nature*. Paris: Flammarion, 1977.

20 op. cit. Kapp.

21 Ibid. Kapp. p. 23.

22 Ibid. Kapp. p. 123.

23 Martin Heidegger (Ed. Krell) *Basic Writings*. London: Routledge & Kegan Paul, 1978.

24 Hölderlin's original is to be found in the poem 'Patmos'.

25 Aristotle, *Nichomachean Ethics*, Book 6, chapters 3–4.

26 Katharina Waack-Erdmann, *Die Demiurgen bei Platon und ihre Technai*. Darmstadt: WBG, 2006.

27 op. cit. Aristotle, Book 6, chapter 4.

28 Max Horkheimer (Trans. Matthew O'Connell), 'Traditional and critical theory', in *Critical Theory*. London: Continuum, 1975. p. 199.

29 Why that might not be true in the non-European or non-Western world is an issue that we will take up in Chapter 6.

30 Christoph Wulf (ed.), *Vom Menschen. Handbuch Historische Anthropologie*. Weinheim: Beltz, 1997.

31 op. cit. Ellul.

32 Gerhard Gamm, 'Technik als Medium. Grundlinien einer Philosophie der Technik', in *Gamm, Nicht nichts. Studien zu einer Semantik des Unbestimmten*. Frankfurt am Main: Suhrkamp, 2000. pp. 275–87. See also: Werner Rammert, 'Technisierung und Medien in Sozialsystemen – Annäherungen an eine soziologische Theorie der Technik', in P. Weingart (ed.), *Technik als sozialer Prozess*. Frankfurt am Main: Suhrkamp, 1989. pp. 128–73.

33 Francis Bacon, 'New Atlantis', in Johnson (ed.), *The Advancement of Learning and New Atlantis*. Oxford: Clarendon Press, 1974.

34 Foucault himself, it might be noted, is looking back to the Aristotelian concept of diaresis, meaning a disposition, arrangement or inner order; already in Aristotle the way that something is shaped by its very possibilities is known as its disposition or diaresis. Vide Aristotle, *Metaphysics*, Book 5, chapter 19 (1022b), 'Disposition is styled an arrangement of that which has parts either according to place or to potentiality, or according to species'.

35 Joseph Weizenbaum, *Die Macht der Computer und die Ohnmacht der Vernunft*. Frankfurt am Main: Suhrkamp, 1977. Hubert L. Dreyfus, *What Computers Can't Do*. New York: MIT Press, 1979. Herbert L. Dreyfus, *On the Internet*. London: Routledge, 2001.

36 op. cit. Böhme and Manzei (eds).

Chapter 2

1 For an overview of the literature, see J. Nordin, Vad är teknik? Filosofiska funderingar kring teknikens Struktur och dynamic, Tema T Rapport 3, 1983, Linköping University.

2 Sybille Krämer, *Technik, Gesellschaft und Natur*. Frankfurt: Campus, 1982. Also Gernot Böhme, 'Die Gesellschaftlichkeit von Natur und Technik'. *Zeitschrift für Hochschuldidaktik* (Sonderheft Nr. 9, Fachübergreifende Lehre an Technischen Universitäten) 8 (1984), 10–26.

3 For a history of the continuities and discontinuities between science and technology, see Gernot Böhme, Wolfgang van den Daele and Wolfgang Krohn, 'Die Verwissentschaftlichung von Technologie', in *Starnberger Studien I*. Frankfurt: Suhrkamp, 1977.

4 Daniel Bell, *The Coming of Post-Industrial Society*. New York: Harper Colophon Books, 1974. Alvin Gouldner, *The Future of Intellectuals and the New Class*. New York: Seabury Press, 1979.

5 For further discussion, see P. Weingart, 'Verwissenschaftlichung der Gesellschaft – Politisierung der Wissenschaft'. *Zeitschrift für Soziologie* 12 (1983), 225–41, and my own book *Ethik leiblicher Existenz*. Frankfurt am Main: Suhrkamp, 2008.

6 See my *Leibsein als Aufgabe: Leibphilosophie in pragmatische Hinsicht*. Zug/ Kusterdingen: Die Graue Edition, 2003.

7 op. cit. Elias.

8 Erik Erikson, *Identität und Lebenszyklus*. Frankfurt: Suhrkamp, 1966.

9 H. Linde, 'Soziale Implikationen technischer Geräte, ihrer Entstehung und Verwendung', in Jokisch (ed.), *Technosoziologie*. Frankfurt: Suhrkamp, 1982.

10 op. cit. Ellul.

11 See my essay 'Quantifizierung und Instrumentenentwicklung. Zur Beziehung der Entwicklung wissenschaftlicher Begriffsbildung und Meßtechnik'. *Technikgeschichte* 43 (1976), 307–13. Also Gernot Böhme and Wolfgang van den Daele, 'Erfahrung als Programm. Über Strukturen vorparadigmatischer Wissenschaft', in *Böhme, van den Daele, Krohn, Experimentelle Philosophie*. Frankfurt: Suhrkamp, 1977. pp. 183–236.

12 This view is set out in Gernot Böhme, 'The knowledge structure of society', in Bergendal (ed.), *Knowledge Policies and the Traditions of Higher Education*. Stockholm: Almquist & Wiksell International, 1984. pp. 5–17.

13 Lewis Mumford, *The Myth of the Machine. Technics and Human Development*. London: Secker & Warburg, 1967.

14 Böhme and Stehr (eds), *The Knowledge Society*. Boston, Dordrecht: Reidel, 1986.

15 My coinage. The term *Technostruktur* was first introduced by me in 'Die Gesellschaftlichkeit von Technik und Natur', in *Zeitschrift für Hochschuldidaktik* 8 (Sonderheft 9). Vienna: Österreichische Gesellschaft für Hochschuldidaktik, 1984. pp. 10–26.

16 Otto Ulrich, Weltniveau. *In der Sackgasse des Industriesystems*. Berlin: Rotbuch, 1979.

17 op. cit. Elias.

18 op. cit. Linde, p. 29.

19 This was Habermas' original point of departure in *Wissenschaft und Technologie als 'Ideologie'*. Frankfurt: Suhrkamp, 1968.

20 Jürgen Habermas, *Theorie des kommunikativen Handelns*. Frankfurt: Suhrkamp, 1981. (Jürgen Habermas (Trans. McCarthy), *The Theory of Communicative Action*. Cambridge: Polity, 1984–7.)

21 Ibid. Vol. 2, p. 179.

22 Ibid. Vol. 2, p. 275.

23 Ibid. Vol. 2, p. 274.

24 Love is represented as this kind of subversive force in Orwell's *1984*.

25 Gogol paints the classic picture of how human beings can continue to exist socially after their physical death in Dead Souls. The social existence of the Russian serfs of the pre-1861 period consisted in their registration as the property of their owners; however, as registration only took place every five years, 'dead souls' continued to have a social half-life until the next register was completed. Today, it is possible to have a similar social half-life as long as one exists as a functional point of connection in technical systems.

26 op. cit. Habermas, p. 273: 'Die Lebenswelt wird für die Koordinierung von Handlungen nicht länger benötigt'.

27 An example will help to convey the point in less abbreviated terms: I can stroll up and down the aisle of a plane, or remain seated (make use of, or abstain from making use of, my body for the purposes of movement) and, in the process, change my position from, say, one continent to the next (traffic).

28 It is important to bear in mind in the latter case that the possibilities opened up by technological development can be closed off again when free activity is again subject to processes of technical rationalization and control. A prime example is contemporary sport.

29 op. cit. Ellul.

30 Lewis Mumford, *Techniques and Civilisation*. New York, London: Harcourt Brace Jovanovich, 1982. Also Lewis Mumford, *The Myth of the Machine. Technics and Human Development*. London: Secker & Warburg, 1967.

31 For detailed studies see, Gernot Böhme, 'The knowledge structure of society', in G. Bergendal (ed.), *Knowledge Policies and the Traditions of Higher Education*. Stockholm: Almquist & Wiksell International, 1984. pp. 5–17. See also *Social Sciences Information* 36 (3) (1997), 447–68, op. cit. Böhme and Stehr.

32 Radovan Richter et al., *Civilisation at the Crossroads: Social Implications of the Scientific and Technological Revolution*. White Plains, NY: International Arts and Sciences Press, 1971.

33 I have discussed the issue at length, critically under the title of modernity's Baconian faith. Vide Gernot Böhme, *Am Ende des Baconschen Zeitalters. Studien zur Wissenschaftentwicklung*. Frankfurt am Main: Suhrkamp, 1993.

34 Alvin Gouldner, *The Future of Intellectuals and the New Class*. New York: Seabury Press, 1979. Georg Konrád and Ivan Szelényi (Trans. Arato and Allen), *Intellectuals on the Road to Class Power*. Brighton, UK: Harvester Press, 1979.

35 Daniel Bell, *The Coming of Post-Industrial Society*. New York: Harper Colophon Books, 1974.

36 See Derek de Solla-Price, *Little Science, Big Science*. New York: Columbia University Press, 1963.

37 Nicolas Rescher, *Scientific Progess*. Oxford: Basil Blackwell, 1978.

38 op. cit. Bell.

39 Rolf Kreibich, *Die Wissenschaftsgesellschaft*. Frankfurt am Main: Suhrkamp, 1986.

40 Florian Znaniecki, *The Social Role of the Man of Knowledge*. New York: Transaction Publishers, 1986.

41 Immanuel Kant, 'An answer to the Question: What is Enlightenment?' (1784). See: http://www.english.upenn.edu/~mgamer/Etexts/kant.html.

42 op. cit. Erikson.

43 Anthony Giddens, *The Consequences of Modernity*. Cambridge: Polity Press, 1991.

44 For an exposition of my notion of the 'souveräner Mensch', see Gernot Böhme, *Anthropologie in pragmatischer Hinsicht*. Frankfurt am Main: Suhrkamp, 1994. Also: 'Souveränität und die Ethik des Pathischen', in Gernot Böhme, *Ethik leiblicher Existenz*. Frankfurt am Main: Suhrkamp 2008.

45 See Plato's dialogue *Laches*.

46 Max Weber, 'Über einige Kategorien der verstehenden Soziologie', in *Gesammelte Aufsätze zur Wissenschaftslehre*. Tübingen: Mohr, 1968.

47 'die Frage nach der Erhaltung des Bestandes von Handlungssystemen' – Niklas Luhmann, *Vertrauen*. Stuttgart: Ferdinand Enke, 1968. p. 2.

48 Cambridge, MA: Riverside Press, 1962.

49 The global financial crisis and its aftermath have brought home the point in a particularly dramatic fashion.

50 Gerald Wagner, 'Vertrauen in Ethik', *Zeitschrift für Soziologie* 23 (1994), 145–57.

51 Ibid.

52 Gernot Böhme, Wolfgang van den Daele and Wolfgang Krohn, 'Die Finalisierung der Wissenschaft', *Zeitschrift für Soziologie* 2 (1973), 128ff.

53 Notable drafts date from 1830 and 1849.

54 One could almost say that collective organization is the very meaning of scientific research as it is conducted in the present era of scientific history.

55 For further details, see my essay 'Schützt das Grundgesetz die Rüstungsforschung?' in Nickel, Roßnagel and Schlink (eds), *Die Freiheit und die Macht – Wissenschaft in Ernstfall*. Baden-Baden: Nomos Verlagsgesellschaft, 1994. pp. 85–92.

56 Again, such a formalistic defence of unlimited freedom of research comes across even more bizarrely when one considers that the individual's opportunities to conduct research are limited by the type of research being conducted within the institution that

employs him. Specifically, younger generations of researchers are not free to choose their own research topics; undergraduate and postgraduate alike have to be integrated into a collective research programme.

57   See Gernot Böhme, William LaFleur, Susumu Shimazono, *Fragwürdige Medizin. Unmoralische medizinische Forschung in Deutschland, Japan und den USA im 20. Jahrhundert*. Frankfurt am Main: Campus, 2008.

58   For a discussion, see *Kommentar zum Grundgesetz für die Bundesrepublik Deutschland in 2 Bänden*. Darmstadt: Luchterhand, 1989. Volume 1, p. 293.

59   Jürgen Habermas, *The Future of Human Nature*. Malden MA: Polity Press, 2003.

60   Ibid.

61   Of course, one remains free to define dignity not just as a right, but also as a duty, or, in the language of Luhmann's *Grundrechte als Institution* (1965) – as an accomplishment. The latter definition is clearly not a concept that can replace the concept of dignity as a natural endowment of human beings, but rather an imperative addressed to fully competent adult human beings. For further discussion, see *Kommentar zum Grundgesetz für die Bundesrepublik Deutschland in 2 Bänden*. Darmstadt: Luchterhand, 1989. Volume 1, p. 204ff.

62   op. cit. Habermas, p. 68.

63   Karl Jaspers, *Philosophie*, Volume 2. Berlin, Göttingen/Heidelberg: Springer, 1956.

64   Hermann Schmitz would speak of a 'subjective Tatsache' – a subjective fact.

Chapter 3

1   Tinguely's machines would be an example.

2   Marx's remarks about technology are scattered throughout his major works. A. A. Kusin gives an impressive overview in *Karl Marx und Probleme der Technik* (Leipzig: VIB Fachbuchverlag, 1970), though his presentation is still clearly shot through with hopes for a bright socialist future.

3   Compare op. cit. Gehlen. Also: Gehlen, *Die Seele im technischen Zeitalter*. Reinbek: Rowohlt, 1957.

4   See Plato's dialogue *Protagoras*, 320d ff.

5   Salomon de Caus, *Von gewaltsamen Bewegungen: Beschreibung etliche so wol nützlichen alß lustigen Machiner* (Frankfurt am Main, 1615), substantially reproduced in 1977 (Hannover: Curt R. Vincentz Verlag).

6   See 'Heron', *Der Kleine Pauly*. Munich: DTV, 1978.

7   C.f. Georg Agricola, *Zwölf Bücher vom Berg- und Hüttenwesen* (Basel, 1556), reprinted Düsseldorf: VDI-Verlag, 1978.

8   op. cit. De Caus. p. 2, preface.

9   One of the few indications that something similar was true in antiquity of the Alexandrian court is to be found in Wolfgang König's discussion of 'Die

Automatentechnik in Alexandria', in *Propyläen Technikgeschichte*. Berlin: Propyläen-Verlag, 1999, Volume 1, p. 202ff.

10 C.f. Alex Sutter, *Göttliche Maschinen. Die Automaten für Lebendiges*. Frankfurt am Main: Athenäum, 1988.

11 C.f. Friedrich Klemm, *Geschichte der naturwissenschaftlichen und technischen Museen*. Munich: R. Oldenburg, 1973. Horst Bredekamp, *Die Geschichte der Kunstkammer und die Zukunft der Kunstgeschichte*. Berlin: Wagenbach, 1993.

12 C.f. Gernot Böhme, 'Kant's epistemology as a theory of alienated knowledge', in Butt (ed.), *Kant's Philosophy of Physical Science*. Dordrecht: Reidel, 1986. pp. 335–50.

13 The relationship was first noted by Blumenberg. See Hans Blumenberg, 'Der Prozeß der theoretischen Neugierde', in *Die Legitimität der Neuzeit*, part 3. Frankfurt am Main: Suhrkamp, 1966. In his efforts to validate modernity, Blumenberg, it must be said, overlooks the effects of courtly curiositas both on technological products and on relations with the natural world.

14 C.f. Fritz Kraft, *Otto von Guericke*. Darmstadt: WBG, 1978, reprinted in 'Die Welt im leeren Raum, Otto von Guericke, 1602–86', exhibition catalogue, Munich: Deutscher Kunstverlag, 2002.

15 C.f. P. Gasparis Schotti e Societate Jesu, *Technica Curiosa, sive Mirabilia Artis* (Nürnberg, 1664), reprinted Hildesheim: Olms, 1977.

16 There is certainly no denying that Guericke understood his scientific enterprise in classic modern terms in opposition to the prescientific book learning of the Middle Ages.

17 '. . . nihil quod Hominis excellentiam, ingenium, industriam extollit altius, quam stupenda Artis technasmata, ab eodem Homine in generis sui seu utilitatem, seu delectationem efformat . . .' Praeloquium ad lectorem, op. cit. Schott, p. 205.

18 op. cit. König, p. 205.

19 C.f. Ulrich Troitzsch, 'Techischer Wandel in Staat und Gesellschaft zwischen 1600 und 1750', in *Propyläen Technikgeschichte* op. cit. Volume 3, p. 39, p. 41.

20 According to Krafft, Guericke's means were limited to 1,000 thalers per year. op. cit. Krafft, p. 10f.

21 Thorstein Veblen, *The Theory of the Leisure Class. An Economic Study of Institutions*. London: Allen and Unwin, 1970.

22 The book appeared in 1913 under the title *Luxus und Kapitalismus*. The full title of later editions is Liebe, *Luxus und Kapitalismus. Über die Entstehung der modernen Welt aus dem Geist der Verschwendung*. Berlin: Wagenbach, 1992.

23 See Thomas Wex, 'Ökonomik der Verschwendung. Allgemeine Ökonomie und die Wirtschaftswissenschaft', in Andreas Hetzel, Peter Wickens and George Bataille, *Vorreden zur Überschreitung*. Würzburg: Königshausen & Neumann, 1999.

24 Walter Benjaminm, we note, thought that the Paris of the high capitalist fin de siècle was among the first historical societies to exemplify the relevant economic/cultural dynamics.

25 C.f. Gernot Böhme, 'Zur Kritik der ästhetischen Ökonomie', in *Zeitschrift für kritische Theorie*, 12/2001, pp. 69–82.

26   Baudrillard already uses these terms in his *La société de consummation, ses myths, ses structures* (Paris: Denoel, 1970), a book that represents an important step towards a post-Marxist analysis of capitalism.

27   C.f. Herbert Marcuse, *Triebstruktur und Gesellschaft*. Frankfurt am Main: Suhrkamp, 1978.

28   The masters thesis of Torsten Dubberke contains a discussion. Dubberke, 'Ich bin ein Tamagotchi. Ein neuer Typ der Beziehung zu technischen Gegenständen'. Darmstadt: Fb 2, 1988. Also www.virtualpet.com.

29   My account will be based on the two volume German edition: Victor Klemperer (Ed. Nowojski and Löser), *Leben sammeln, nicht fragen wozu und warum (1918–32)*. Berlin: Aufbau-Verlag, 1996; Klemperer (Ed. Nowojski and Hedwig Klemperer) *Ich will Zeugnis ablegen bis zum letzten (1933-45)*. Berlin: Aufbau-Verlag, 1995. For an abridged English version of the latter volume, see Klemperer (Trans. Chambers), *I Shall Bear Witness: Diaries 1933–41*. London: Weidenfeld and Nicholson, 1998 and Klemperer (Trans. Chambers), *To the Bitter End. The Diaries of Victor Klemperer 1942–5*. London: Weidenfeld and Nicholson, 1999. Translations of Klemperer in the present chapter are mine (CS).

30   Klemperer uses the form of racist disparagement that was conventional in his day.

31   Klemperer (Trans. Chalmers), *I Shall Bear Witness. The Diaries of Victor Klemperer 1933–41*. London: Phoenix, 1999, p. 236.

32   Ibid. pp. 340–1.

Chapter 4

1   For a closer examination of the distinction: Gernot Böhme, 'Naturwissenschaft der Technik oder die Frage nach einem neuen Naturbegriff'. *Technik kontrovers* 3 (1982), 62–70.

2   Gernot Böhme, 'Technologiekritik als gesellschaftlicher Konflikt', in F. Moser (ed.), *Neue Funktionen von Wissenschaft und Technik in den 80er Jahren*. Vienna: Verlag der wissenschaftlichen Gesellschaft Österreichs, 1981, pp. 52–76.

3   Serge Moscovici, *Essai sur l'histoire de la nature*. Paris: Flammarion, 1977.

4   Moscovici's scheme for describing the relationship between nature and technology, we note, remains silent on the question of where man actually comes upon the forms that guide his modelling (remodelling) of nature. For the longest time, from antiquity until the eighteenth century, technological activity was understood as an imitation of nature; indeed, in the nineteenth century, Ernst Kapp still busied himself interpreting technology as an externalization of human bodily processes – what we might call a displacement of the functions of human bodily organs into the world. As shaky an interpretative move as that sounds to us today, it still seems possible, indeed important, to think through another type of emergent property of technology – what takes place when technology gives birth to structures that have gone unrealized in nature itself, viz.

in the fulfilment of social functions. Here, the wheel would be the most obvious case where the organizational forms that technology impresses on nature have their origins in fundamentally social contexts rather than bodily organic contexts.

5 Sybille Krämer, *Technik, Gesellschaft und Natur*. Frankfurt: Campus, 1982.

6 Lewis Mumford, *The Myth of the Machine. Technics and Human Development*. London: Secker & Warburg, 1967.

7 Gernot Böhme, 'Der normative Rahmen wissenschaftlich-techischen Handelns', in *Deutsches Institut für Normierung* (ed.), *Regeln und Normen in Wissenschaft und Technik*. Berlin: Beuth, 1984, pp. 11–20.

8 In Germany, for instance, the relevant codes set down by agencies such as the DIN (Deutsches Institut für Normierung) or VDE (Verband der Elektrotechnik).

9 Over time, indeed, this normative framework of technological action has become so narrow that practising professionals begin to wonder whether innovative engineering is suffering as a result.

10 op. cit. Böhme and Schramm.

11 For more on the notion of socially constituted nature, see Gernot Böhme, 'Was ist sozial konstituierte Natur?' *Öko-Mitteilungen* (March 1983), 27–8.

12 Odo Marquard, *Transzendentaler Idealismus. Romantische Naturphilosophie. Psychoanalyse*. Cologne: Jürgen Dinter, 1987.

13 op. cit. Böhme, 1994.

14 Martin Heidegger (Trans. Macquarrie and Robinson), *Being and Time*. Oxford: Basil Blackwell, 1987. §4, p. 32.

15 Jean-Paul Sartre (Trans. Mairet), 'Existentialism is a Humanism'. http://www.marxists. org/reference/archive/sartre/works/exist/sartre.htm.

16 Christoph Wulf (ed.), *Vom Menschen. Handbuch Historische Anthropologie*. Weinheim: Beltz, 1997.

17 Bruno Snell, *Die Entdeckung des Geistes*. Göttingen: Vandenhoeck & Bouvier, 1986.

18 Hermann Schmitz, *System der Philosophie*. Volume 2, part 1 – *The Body*. Bonn: Bouvier, 1965.

19 For more detail, see op. cit. Böhme, 1994, Böhme, *Leibsein also Aufgabe: Leibphilosophie in pragmatische Hinsicht*. Zug/Kusterdingen: Die Graue Edition, 2003.

20 See chapter 2, 'Civilisation in the Age of Techno-Science'.

21 See Gernot Böhme, 'Die Wirklichkeit der Bilder und ihr Gebrauch', in *Hestia: Jahrbuch der Klages-Gesellschaft*. Würzburg: Königshausen & Neumann. Volume 22 (2004/7), pp. 139–50.

22 This older conception of the imagination as a capacity for configuring the body is beautifully illustrated in Adam Bernd's autobiographical writings – Bernd, *Eigene Lebens-Beschreibung (1738)*. Munich: Winkler, 1973. §§ 65 and 72.

23 My translation (CS). See also: Paul Feyerabend, *Against Method. Outline of an Anarchistic Theory of Knowledge*. London: Verso, 1978.

24 Barbara Duden, *Der Frauenleib als öffentlicher Ort. Vom Mißbrauch des Begriffs Leben*. Hamburg: Luchterhand, 1991, p. 90.

25 op. cit. Krämer (1982).

26 op. cit. Gehlen (1957).

27 op. cit. Gehlen (1962).

28 Rudolf zur Lippe, *Sinnesbewußtsein. Grundlegung einer anthropologischen Ästhetik.* Reinbek: Rowohlt, 1987. Dieter Hoffmann-Axthelm, *Sinnesarbeit. Nachdenken über Wahrnehmung.* Frankfurt am Main: Campus, 1984.

29 Werner Kutschmann, *Der Naturwissenschaftler und sein Körper.* Frankfurt am Main: Suhrkamp, 1986.

30 op. cit. Duden, p. 25.

31 Further details of the distinction are to be found in Hermann Schmitz, 'Der unerschöpfliche Gegenstand', in *Grundzüge der Philosophie.* Bonn: Bouvier, 1990, p. 6ff et passim.

32 op. cit. Duden, p. 25.

33 Ibid. p. 27.

34 Ibid. p. 38f.

35 Eva Schindele, *Gläserne Gebär-Mutter. Vorgeburtliche Diagnostik – Fluch und Segen.* Frankfurt am Main: Fischer, 1990, p. 22.

36 See Anderson (ed.), *Catalogue – Perlon, Petticoat, Pesticide.* Basel/Berlin: Reinhardt, 1994.

37 Shuhei Hosokawa, 'The walkman effect', in Barck, Gente, Paris and Richter (eds), *Aisthesis. Wahrnehmung heute oder Perspectiven einer anderen Ästhetik.* Leipzig: Reclam, 1990. pp. 229–51.

38 Helga Nowotny has given a striking demonstration of the quandary in research focusing on expert submissions to government relating to the development of nuclear energy. See Nowotny, *Kernenergie: Gefahr oder Notwendigkeit.* Frankfurt am Main: Suhrkamp, 1979.

39 'Bioethik als Rechfertigungsmechanismus? Hans-Martin Sass antwortet Kritikern der Bioethik', op. cit. pp. 76–87.

40 See op. cit. Böhme, 2008. In particular, refer to the chapter 'Substantielle Sittlichkeit oder das Übliche'.

41 For these and related definitions of the fundamental problems of ethics, see Gernot Böhme (Trans Jephcott), *Ethics in Context. The Art of Dealing with Serious Questions.* Cambridge: Polity, 2001.

42 My own non-traditional philosophical term for what stands to be integrated here is 'the other of reason'. See: Hartmut Böhme and Gernot Böhme, *Das Andere der Vernunft. Zur Entwicklung von Rationalitätsstrukturen am Beispiel Kants.* Frankfurt am Main: Suhrkamp, 1985.

43 Günther Anders, *Die Antiquiertheit des Menschen*, 2 volumes. Munich: C. H. Beck, 1987.

44 Hannah Arendt, *Vita Activa oder vom tätigen Leben.* Munich: Piper, 1989.

45 The individual liberty to think of oneself as a new beginning is, contra Habermas, not guaranteed by any mere *Recht auf Zufall* ('right to chance'), which must be rejected on

the grounds that it is too external to secure enough depth for the required conception of individuality. An adequate conception, in our opinion, ought to allow the individual to lay claim to being a new beginning, not just a continuation of human life.

46 See chapter 1.5 ('Der Mensch') in Gernot Böhme, *Einführung in die Philosophie*, Frankfurt am Main: Suhrkamp, 2001.

47 For more on the thesis that these three concepts represent the defining traits of self-cultivation in Kant, see Gernot Böhme, 'Was wird aus dem Subjekt? Selbstkultivierung nach Kant', in Kaplow (ed.), *Nach Kant: Erbe und Kritik*. Münster: LIT Verlag, 2005. pp. 1–16.

Chapter 5

1 My translation (CS). For a complete English version of Sulzer, see Sulzer (Trans. Miller), *Dialogues on the Beauty of Nature and Moral Reflections on Certain Topics of Natural History*. Lanham: University of America Press, 2005.

2 Aristotle, *Physics*. B8, 199a.

3 H. Koller, *Die Mimesis in der Antike. Nachahmung, Darstellung, Ausdruck*. Bern: A. Francke, 1954.

4 G. Bien, 'Bemerkungen zu Genesis und ursprünglicher Funktion des Theorems von der Kunst als Nachahumng der Natur'. *Zeitschrift für Literatur, Kunst und Philosophie*, no. 2. Münster, 1964, 26–43.

5 Aristotle, *Physics*. B1 (Book 2).

6 A. Sutter, *Göttliche Maschinen. Die Automaten für Lebendiges bei Descartes, Leibniz, La Mettrie und Kant*. Frankfurt am Main: Athenäum, 1988. D'Alembert's article 'Automates' in the Encyclopédie is particularly informative on the subject of Vaucanson's duck.

7 op. cit. Kapp.

8 Gehlen, *Die Seele im technischen Zeitalter*. Reinbek: Rowohlt, 1957.

9 Manfred Schenker, *Charles Batteux und seine Nachahmungstheorie in Deutschland*. Leipzig: Haessel, 1909; Irmela von Lühe, *Natur und Nachahmung. Untersuchungen zur Batteux-Rezeption in Deutschland*. Bonn: Bouvier, 1979.

10 Fritz Krafft, 'Die Stellung der Technik zur Naturwissenschaft in Antike und Neuzeit'. *Zt [GBGB] Technikgeschichte* 37 (3) (1970), 189–209.

11 Stephen Toulmin, *Foresight and Understanding: An Enquiry into the Aims of Science*. London: Hutchinson, 1961.

12 Sybille Krämer, *Technik, Gesellschaft und Natur*. Frankfurt am Main: Campus, 1987.

13 Hans Blumenberg, 'Nachahmung der Natur: Zur Vorgeschichte der Idee des schöpferischen Menschen'. *Studium Generale* 10 (1957), 266–83.

14 Ibid. p. 268.

15 Quoted in W. Haftmann, *Wege bildnerischen Denkens*. Frankfurt am Main: Fischer, 1961, p. 71. Unfortunately, Haftmann fails to specify his sources for his Klee citations. His work, it should be noted, also shows that Klee is in no sense a straightforward mimetoclast – that he, in certain senses, also stands close to the mimetic thesis.

16 My italics (CS). Our citation provides the first line of Klee's contribution to the collection entitled 'Schöpferische Konfession' (Creative Confession) in the catalogue (Ed. Güse) Paul Klee, *Wachstum regt sich. Klees Zwiesprache mit der Natur*. Munich: Prestel, 1990. p. 57.

17 S. Koppelkamm, *Künstliche Paradiese, Gewächshäuser und Wintergärten des 19. Jahrhunderts*. Berlin: Ernst, 1988.

18 'A lot of anxiousness and hostility against AI is motivated by the fantastic speculations about so-called reasoning machines which are not technologically justified in any way'. K. Mainzer, 'Knowledge-based systems'. *Zeitschrift für allgemeine Wissenschaftstheorie* 21 (1990), 48.

19 H. Lübbe, 'Was heisst: "Das kann man nur historisch erklären"?', in Hübner and Menne (eds), *Natur und Geschichte, 10. Deutscher Kongress für Philosophie, Kiel, October, 1972*. Hamburg: Meiner, 1973, pp. 207–16.

20 See chapter 3, 'Die Geste der Natürlichkeit' in op. cit. Böhme, 1997.

21 Immanuel Kant, *Critique of Judgment*, §45.

22 See Gernot Böhme, 'Wissenschaftliches und lebensweltliches Wissen am Beispiel der Verwissenschaftlichung der Geburtshilfe', in *Gernot Böhme, Alternativen der Wissenschaft*. Frankfurt am Main: Suhrkamp, 1993.

23 G. Corea, *Mutter Maschine. Reproduktionstechnologien – von der künstlichen Befruchtung zur künstlichen Gebärmutter*. Berlin: Rotbuch Verlag, 1986. Also: Bradisch, Feyerabend and Winkler (eds), *Frauen gegen Gen- und Reproduktionstechnologie*. Munich, 1989. For a contrasting opinion: F. Akashe-Böhme, 'Selbstbestimmung an den Grenzen der Natur', in Konnertz (ed.), *Grenzen der Moral. Ansätze feministischer Vernunftkritik*. Tübingen: Konkursbuch, 1991. pp. 13–29.

24 Otto Ulrich, *Technik und Herrschaft*. Frankfurt am Main: Suhrkamp, 1977.

25 Ulrich Beck, *Risk Society: Towards a New Modernity*. London: Sage, 1992.

26 Joseph Weizenbaum, *Die Macht der Computer und die Ohnmacht der Vernunft*. Frankfurt am Main: Suhrkamp, 1978.

27 C.f. W. Schäfer, 'Die Büchse der Pandora. Über Hans Jonas, Technik, Ethik und die Träume der Vernunft'. *Merkur* 34 (1989), 292–304.

28 In Walter Benjamin (Trans. Zohn), *Illuminations*. New York: Schocken Books, 1969. pp. 217–52.

29 For more on the Benjaminian concept of aura, see Marleen Stoessel, *Aura. Das vergessene Menschliche. Zur Sprache und Erfahrung bei Walter Benjamin*. Munich: Hanser, 1983.

30 Martin Heidegger, *Beiträge zur Philosophie (Vom Ereignis)*. Collected Works, Volume 65, Frankfurt am Main: Klostermann, 1989. p. 21f. and passim.

31 For more on this thesis, see op. cit. Böhme and Schramm (eds).

32 One of the main proponents of the term 'Mensch-organisierte Ökosysteme' is H. Sukopp, long-standing Professor of Ecology at Berlin's Technical University.

33 op. cit. Böhme and Schramm.

34 Again, see chapter 3, 'Die Geste der Natürlichkeit', of op. cit. Böhme, 1997.

35 The term was coined by Michael Pollan. See, for example, *The Omnivore's Dilemma. The Natural History of Four Meals*. New York: Penguin Press, 2006.

36 For more on anthropological changes brought about by technology, see my *Anthropologie in pragmatischer Hinsicht* (Frankfurt am Main: Suhrkamp, 1994) and the essay 'Anthropological Change in Technological Civilisation' in the present volume.

37 Aristotle, *Physics*. B1.

38 Immanuel Kant, *Critique of Judgment*, §45.

39 Ibid. B77.

40 The tension becomes most evident in experiments conducted on animals.

41 op. cit. Kant, §42.

42 Facticity and project – the terminology stems from Heidegger's *Being and Time* (1926).

43 'I am my body and at the same time have a body' – a basic condition of human existence dubbed *Exzentrizität* (eccentricity) by H. Plessner.

44 Wolfgang van den Daele, *Mensch nach Maß*. Munich: Hanser, 1985.

45 One's own natural being itself becomes a moral problem. For more, see op. cit. Böhme, 2008.

46 At least not in Benjamin's sense of culture, which was still essentially that of the educated middle classes of the European nineteenth and twentieth centuries. The situation looks different if one brings pop-culture into consideration.

47 See 'NS-Architektur als Kommunikations-Design' in Gernot Böhme, *Architektur und Atmosphare*. Munich: Wilhelm Fink Verlag, 2006.

48 For more on the notion of socially constituted nature, see op. cit. Böhme and Schramm, 1985.

Chapter 6

1 Max Horkheimer (Trans. Matthew O'Connell), 'Traditional and critical theory', in *Critical Theory*. London: Continuum, 1975.

2 'Philosophie und Kritische Theorie', in Herbert Marcuse, *Kultur und Gesellschaft I*. Frankfurt am Main: Suhrkamp, 1968.

3 An indication of my own work along these lines is to be found in *Einführung in die Philosophie. Weltweisheit, Lebensform, Wissenschaft*. Frankfurt am Main: Suhrkamp, 2001, chapter 3.9.

4 My translation. (CS)

5 Theodor W. Adorno et al., *Der Positivismusstreit in der deutschen Soziologie*. Neuwied, Berlin: Luchterhand, 1971.

6   Alfred Schmidt, *Der Begriff der Natur in der Lehre von Marx*. Frankfurt am Main: EVA, 1974.

7   op. cit. Böhme and Schramm (eds).

8   op. cit. Horkheimer, p. 219.

9   Ibid. p. 215.

10 Ibid. p. 218.

11 Wolfgang Welsch, *Vernunft. Die zeitgenössische Vernunftkritik und das Konzept der transversalen Vernunft*. Frankfurt am Main: Fischer, 1985.

12 Jürgen Habermas, 'Die neue Unübersichtlichkeit', in *Die Krise des Wohlfahrtstaats und die Erschöpfung utopischer Energien*. Frankfurt am Main: Suhrkamp, 1987. p. 146.

13 op. cit. Marcuse, p. 121.

14 Friedrich Nietzsche, *The Gay Science*, No. 125.

15 See also Jürgen Habermas, 'Die Einheit der Vernunft in der Vielfalt der Stimmen', in *Nachmetaphysisches Denken. Philosophische Aufsätze*. Frankfurt am Main: Suhrkamp, 1988. pp. 153–86. Welsch addresses this line of Habermas' thought on p. 756 of op. cit. Welsch.

16 Ulrich Pothast, *Lebendige Vernünftigkeit. Zur Vorbereitung eines menschenangemessenen Konzepts*. Frankfurt am Main: Suhrkamp, 1998.

17 Herbert Schnädelbach, *Zur Rehabilitierung des animale rationale. Vorträge und Abhandlungen 2*. Frankfurt am Main: Suhrkamp, 1992. p. 47f.

18 The allusion is to Vattimo's notion of a pensiero debole. See op. cit. Welsch, particularly p. 826f.

19 Robert B. Brandom, *Making It Explicit. Reasoning, Representing and Discursive Commitment*. Cambridge, MA: Harvard University Press, 1994.

20 See Gernot Böhme, 'Kritische Theorie der Natur'. *Zeitschrift für kritische Theorie 9* (1999), 59–71.

21 op. cit. Böhme (2003).

22 op. cit. Böhme and Schramm.

23 op. cit. Böhme and Stehr.

24 Gernot Böhme, 'The structures and propects of knowledge society'. *Social Sciences Information 36* (3) (1997), 447–68.

25 The thesis that Plato's dispute with the sophists was already a struggle for power within the field of higher education is something I have attempted to demonstrate in an essay, 'Demarcation as a Strategy of Exclusion: Philosophers and Sophists' in op. cit. Böhme and Stehr.

26 P. Janich, 'Informationsbegriff und methodische kulturalistische Philosophie'. *Ethik und Sozialwissenschaft 9* (1998), 169–82.

27 The societies in questions are, incidentally, those in which the politics of science (science policy etc.) is explicitly formulated and, sometimes, openly contested.

28 For more on the terminological distinction I am making here between Realität and Wirklichkeit, see Gernot Böhme, *Theorie des Bildes*. Munich: Fink, 1999.

29   Sartre gives the problem a literary form in Nausea through the character of the
     autodidact who attempts to educate himself by reading the contents of the library
     in alphabetical order.

30   Wilfried Brauer, Wolfhart Haacke and Siegfried Münch, *Studien- und Forschungsführer
     Informatik*. Heidelberg: Springer, 1984. p. 34.

31   Alfred Lothar Luft, *Informatik als Technikwissenschaft. Eine Orientierungshilfe für das
     Informatikstudium*. Mannheim, Vienna, Zürich: BI Wissenschaftsverlag, 1988. p. 14.
     Luft himself refers to the work of H. Zemanek.

32   'Schulen ans Netz, Ergebnisse und Perspektiven'. *Computer und Unterricht* 41
     (1.2001), 10.

33   Ibid. p. 8.

34   Don Tapscott, *Growing Up Digital. The Rise of the Net Generation*. New York:
     McGraw Hill, 1988. p. 138.

35   Burkhardt Strassmann, 'Lernen mit dem Computer. Wie die neuen Medien Schule und
     Hochschule erobern'. *Zeitpunkte* 1 (2000), 28.

36   *Computer und Unterricht* (op. cit.), 6.

37   *Zeitpunkte* 1 (2000), 22. For comparison with the US position, see Cordes and
     Miller (eds), *Fool's Gold: A Critical Look at Computers in Childhood*, http://www.
     allianceforchildhood.net. According to the estimates set out in the text, government
     spending on computerization in schools for the financial year 1999–2000 was in
     the order of US$7.8 billion. An advisory committee established by President Clinton
     recommended that this already hefty sum be raised to US$15 billion.

38   *Süddeutsche Zeitung* (19/10/1999).

39   *Stuttgarter Zeitung* (19/6/2001).

40   According to the Bertelmann Foundation's Head of Media Research, Ingrid Hamm.
     See *Zeitpunkte* 1 (2000), 23.

41   *Darmstädter Echo* (6/11/1999).

42   op. cit. *Stuttgarter Zeitung*.

43   *Computer und Unterricht* 41, 18.

44   Ibid. p. 7.

45   Ibid. p. 10.

46   C.f. op. cit. Cordes and Miller, chapter 1, 'Healthy children: Lessons from research on
     childhood development'. The studies of J. Piaget are also relevant.

47   Nico Stehr and Reiner Grundmann, 'Die Arbeitswelt in der Wissensgesellfschaft', in
     Kurtz (ed.), *Aspekte des Berufs in der Moderne*. Opladen: Leske und Budrich, 2001.

48   For different versions of the same argument, see Gernot Böhme, 'Bildung als
     Widerstand', in *Die Zeit* (16/9/1999); 'Antizyklisch denken', in *Argauer Zeitung*
     (29.4.2000); Heike Schmoll, 'Nur wer lessen kann, wird den Computer beherrschen', *in
     Frankfurter Allgemeiner Zeitung* (22/4/2001); 'Wer klassisch gebildet ist, surft besser', in
     Badische Zeitung (5/10/2000).

49   op. cit. Böhme and Stehr.

50 Gernot Böhme, 'Wie natürlich ist die natürliche Geburt?', in op. cit. Böhme, 1997.

51 op. cit. Tapscott, 1998.

52 Bernard de Mandeville, *Treatise of the Hypochondriack and Hysteric Diseases in Three Dialogues*. London: J. Tonson, 1730 (second edition).

53 op. cit. Elias. C.f. Gernot Böhme, 'Zivilisation und Humanität. Zur zivilisatorischen Bedeutung von Technik', in Gamm and Kimmerle (eds), *Vorschrift und Autonomie. Zur Zivilisationsgeschichte von Moral*. Tübingen: Edition Discord, 1989, pp. 11–28. Also chapter 11, 'Technische Zivilisation', in op. cit. Böhme, 1994.

54 Maybe this overstates the case slightly. Elias' analysis of the introduction of knife and fork seems to point in the direction of the reading of technology's civilizing function that we have suggested would be fruitful to pursue much further.

55 Kwame Gyekye, *Tradition and Modernity. Philosophical Reflections on the African Experience*. Oxford: Oxford University Press, 1997. p. 280.

56 Gernot Böhme, 'Wie natürlich ist die natürliche Geburt?' in op. cit. Böhme, 1997.

57 Grantly Dick-Read, *Childbirth Without Fear. The Principles and Practice of Natural Childbirth*. London: Pinter & Martin, 2004.

58 For a lengthier discussion, the reader is referred to op. cit. Böhme, 1997, II.5.

59 Jürgen Habermas (Trans. Lawrence), *The Philosophical Discourse of Modernity*. Cambridge, MA: MIT Press, 1987, 2 volumes.

60 Philippe Ariès, *Geschichte der Kindheit*. Munich: Deutscher Taschenbuchverlag, 2003.

61 See Gernot Böhme, 'Die Bedeutung des englischen Gartens und seiner Theorie für die Entwicklung einer ökologischen Naturästhetik', in *Für eine ökologische Naturästhetik*. Frankfurt am Main: Suhrkamp, 1989.

62 The law states, in short, that animals are to be respected as fellow creatures.

63 My analysis of the debate about German Animal Protection Law is to be found in op. cit. Böhme, 2001, p. 120ff.

64 Isaiah, 45.4.

65 R. P. Sieferle, *Fortschrittsfeinde? Opposition gegen Technik und Industrie von der Romantik bis zur Gegenwart*. Munich: C. H. Beck, 1984. See, in particular, chapter 13.

66 Callicott and Ames (eds), *Nature in Asian Traditions of Thought: Essays in Environmental Philosophy*. Albany, NY: State University of New York Press, 1989.

67 For the extremely various meanings of chi in everyday Asian lifeworlds, see Kimura Bin, *Zwischen Mensch und Mensch. Strukturen Japanischer Subjektivität*. Darmstadt: WBG, 1995. pp. 119–31.

68 'Confucianism disclosed its religious dimension in its insistence that dead ancestors were de facto deities and that their descendants' filial piety was expressed in a total acceptance of their own bodies, given them as somatic inheritances. This entailed that any intentional alteration of it – tattooing, piercing, etc. – constituted a violation of religious duty and placed the all-important beneficence of a dead but still involved ancestor's favour in jeopardy'. Willian LaFleur, 'Body' in Taylor (ed.), *Critical Terms for Religious Studies*. Chicago, IL: University of Chicago Press, 1998. p. 39f.

Appendix

1 Friedrich Nietzsche (Trans. Hollingdale), *Thus Spoke Zarathustra*. London: Penguin, 1969. pp. 46–7.

2 Ibid. p. 43.

3 Jean-Luc Nancy, *L'Intrus*, Paris: Galilée, 2000; *Der Eindringling. Das fremde Herz* (bi-lingual). Berlin: Merve, 2000. A preliminary English translation by Susan Hanson exists at http://www.scribd.com/doc/32475844/Nancy-Jean-Luc-L-intrus-English. Unless otherwise marked, translations of Nancy are by CS and refer to the French-German edition.

4 Jean-Luc Nancy, *Der Eindringling. Das fremde Herz* (bi-lingual). Berlin: Merve, 2000. p. 16/18.

5 Ibid. p. 42/6.

6 Ibid. p. 48.

7 The foreigner who 'arrives today and remains tomorrow'. Georg Simmel, *Der Fremde*. Frankfurt am Main: Suhrkamp, 1987. p. 63.

8 op. cit. Nancy. http://www.scribd.com/doc/32475844/Nancy-Jean-Luc-L-intrus-English.

9 op. cit. Nancy. http://www.scribd.com/doc/32475844/Nancy-Jean-Luc-L-intrus-English.

10 Martin Heidegger (Trans. Macquarrie and Robinson), *Being and Time*. Oxford: Basil Blackwell, 1987. p. 95ff.s

11 op. cit. Nancy. http://www.scribd.com/doc/32475844/Nancy-Jean-Luc-L-intrus-English.

12 op. cit. Nancy, p. 32.

13 Adorno and Horkheimer (Trans. Jephcott), 'The concept of enlightenment', in *Dialectic of Enlightenment*. Stanford: Stanford University Press, 2002. chapter 1.

14 op. cit. Nancy, p. 34.

15 The concept stems from Erik H. Erikson. See *Identität und Lebenszyklus*. Frankfurt am Main: Suhrkamp, 1966.

16 op. cit. Nancy, p. 38.

17 Ibid. p. 42.

18 op. cit. Böhme (2001).

19 op. cit. Nancy, p. 20.

20 Plato, *Gorgias*. 511d, 511e.

21 op. cit. Nancy, p. 24, 26.

22 For more on this topic and related terminology, see op. cit. Schmitz, Volume 2, chapter 1, 'The Body'.

INDEX

ff denotes the following page references.